AF378970

LANDSCAPES OF LOSS

'No story of "new" India deserves to be told quite like that of the farmers of Marathwada. It's a tale of grit and pathos, of those whose lives have been reduced to a farm suicide statistic. By unflinchingly reporting on their plight, Kavitha Iyer fills an important gap in contemporary journalism: this is a book that is a much-needed wake-up call, one that gives a face at last to an unfolding human tragedy.'

— RAJDEEP SARDESAI, news anchor, journalist and author

'This is a sad, sad tale of a region so rich with people's love and resistance but plagued by nature's betrayal, sahukar's savagery, landlord's casteism and the government's barbarity. Iyer has written a rigorously researched text that will force you to think about my home – Marathwada – that I carry in me. For Marathwada is a consciousness and Iyer has done a phenomenal job of bringing forth the pain and anguish of this unfortunate region. An essential deposition for policymakers, organizers, students and wannabe politicians. It is a welcome addition to the literature on Marathwada politics.'

— SURAJ YENGDE, scholar and author

'An evocative and nuanced account of everyday tragedy and some triumphs from a desperate region.'

— MEENA MENON, independent journalist and author

'Only 300 to 600 kilometres from Mumbai, Marathwada has remained an orphaned region, neglected by development processes. Kavitha Iyer's captivating and compassionate canvas, aptly titled *Landscapes of Loss*, is an account of the tenacity, grit and life force of the people here – narratives that evoke amazement and agony. Neither the state nor the political class recognizes the struggles of this region's people against an oppressive social hierarchy. Sometimes, these battles end in farmer suicides. Kavitha's explorations are a call for policymakers and strategists to wake up to Marathwada's pain.'

— KUMAR KETKAR, Member of Parliament, journalist and author

'Kavitha Iyer's book identifies crucial issues of agrarian crisis. Although this subject is somewhat emotional, the author has exerted both logic and passion in making the book timely and relevant. A combination of small holdings, rising input costs, uncertainties in climate and the market for farm goods has impaired farming viability. Cyclical drought engulfs every aspect of the village economy. Iyer brings to light that farmer suicides are not just a reflection of the agrarian crisis but a sad manifestation of the water crisis.'

— VINAYAK S. DESHPANDE, Vice Chancellor (Acting), RTM Nagpur University

KAVITHA IYER

LANDSCAPES OF LOSS

THE STORY OF AN INDIAN DROUGHT

HarperCollins *Publishers* India

First published in India by
HarperCollins *Publishers* in 2021
A-75, Sector 57, Noida, Uttar Pradesh 201301, India
www.harpercollins.co.in

2 4 6 8 10 9 7 5 3 1

P-ISBN: 978-93-9032-746-1
E-ISBN: 978-93-9032-747-8

Typeset in 11/15.2 Berling LT Std at
Manipal Technologies Limited, Manipal

Printed and bound at
Thomson Press (India) Ltd.

This book is produced from independently certified FSC™ paper
to ensure responsible forest management.

For India's *annapurnas* and *annadatas* –
the farmers we have failed.

CONTENTS

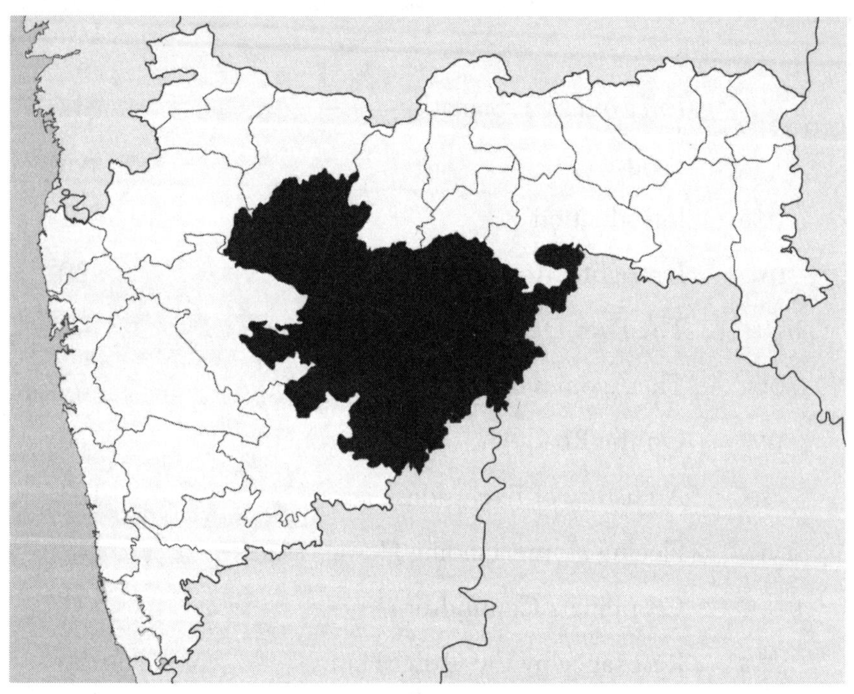

Map of Maharashtra with the Marathwada region highlighted
Source: Wikimedia Commons

FOREWORD

～

S HE HAD PAID a full rupee for a litre of water and had picked up some fifteen litres. That wasn't the regular price but an extraordinary one, she patiently explained, seeing our startled expressions. At the start of the scarcity season in this Aurangabad area, Shantabai was paying 20–25 paise a litre. But now, with scarcity at a peak, she stood for hours in the queue and paid whatever was demanded. 'The public taps are dry,' she said.

Considering what the beer and alcohol factories were paying for water in the same region at the same time, that was quite a price. There were about ten of them at the time, in 2016, and they were paying 4 paise a litre for 3–5 million litres daily. They had, in fact, paid 1 paisa a litre for ages, before a non-governmental organization took the matter to the courts. Anyway, the issue was settled with them having to pay 'four times more'. Yes, 4 paise.

Consider this: the highest – extortionist – rate that the poor agricultural worker Shantabai paid was twenty-five times higher than the standard rate the alcohol industry enjoyed for nearly two decades. The 'standard' summer rate she said she normally paid was still five times greater than the highest rate the beer factories paid in 2016.

Public outrage over the grossly unequal access to water flared up across Maharashtra that year. The Indian Premier League was forced to shift games scheduled in the state to venues outside it. (Stadiums hosting major sporting events consume massive quantities of water.) The Aurangabad bench of the Bombay High Court called for a 60 per cent cut in water to liquor factories in twelve districts of Maharashtra. Buildings coming up with a swimming pool on every floor – 'balcony pools' – were compelled to shelve that part of their plans, at least for a while. And after further protests, even the beer factories had to pay – one and a half paisa more. The rate for them, since 2018, is 5.5 paise per litre. I'm sure Shantabai is still paying the rates she did, or more.

Marathwada is painful proof that drought is not simply a natural calamity, that it is so largely driven by human agency and commercial greed. That insatiable plunder for profit has given rise to a thriving 'thirst economy'. There are some months in the year when two sectors of Maharashtra's thirst economy make profits that would stoke envy in Dalal Street and the many chambers-of-commerce fraternities: the borewell industry and the tanker industry.

It was in 2016 that I ran into new owners of borewell rigs (almost all of which come from a single town in Tamil Nadu – Tiruchengode in Namakkal district) on the highways and backroads of Marathwada. One of them was pleased to share the information that he had bought a rig for Rs 1.4 crore – and had recovered most of its cost in four months and would clear all remaining dues in another two. Farmers at the time were spending around Rs 1,50,000 in sinking a 500-foot borewell.

Three years earlier, I had come across the village of Takwiki in the Osmanabad district of Marathwada. There had been close to 1,500 borewells sunk in the village in just a few years. Most of them had failed. Between January and March alone, Takwiki had seen nearly 300 new bore points sunk in desperation to varying depths.

Farmer desperation rises in inverse proportion to both monsoon and groundwater levels. Thousands of borewells actually fail. Yet, those whose borewells fail often take yet another loan and sink a couple more. With interest rates ranging from 60 to 120 per cent there, an explosion of debt is the most predictable outcome. And borewell bankruptcies are many. They have even been a factor in several farmer suicides in Marathwada.

Meanwhile, in the town of Jalna alone, tanker operators selling water were touching sales of close to Rs 1 crore a day by 2016. All these rates have only gone up since then, but so have the profits reaped from exploiting scarcity.

The unchecked guzzling of groundwater in the region has even seen a few – but worrying – instances of striking what are called 'paleo-historic storages' as the wells go deeper. That is, water which is many millennia old.

The largely – in practice – unregulated loot of groundwater saw officials of the Groundwater Surveys and Development Agency author reports in which they raised red flags. But in Marathwada, as always, those flags flew at half-mast.

Water is more often than not the single most explosive political issue in this country. The greatest inter-state political conflict in the south of India is the unending Cauvery water dispute between Karnataka and Tamil Nadu. There are also other less high-profile but very serious ones: the Krishna water dispute between Andhra Pradesh, Karnataka and Maharashtra has seen temporary solutions. But those won't hold as the water situation changes. The dispute over the Mulaperiyar Dam between Tamil Nadu and Kerala continues to simmer. Kerala and Karnataka have seen conflicts over the Kabini River.

A similar picture prevails, just less seen, across the rest of the country too. With Andhra Pradesh and Telangana now different states, all the water-sharing agreements and awards involving the formerly unified state have become twice as complicated. A generation has grown up with little idea of how the Khalistan movement in Punjab was able to capitalize on grievances over the sharing of Ravi and Beas river waters between the states of Punjab, Rajasthan and Haryana.

And then there's the question of how we use the water we do have. In the drought-prone parts of Marathwada region, we grow (among other things) lots of sugarcane, one of the greatest water guzzlers imaginable.

~

MARATHWADA'S LARGELY CONSTRUCTED drought was a book that needed writing. Kavitha Iyer has now written it. And hers is an important book indeed. One based on very many years of meticulous field reporting, painstaking research and much empathy.

Kavitha does understand that the agrarian crisis in Marathwada and Maharashtra 'is only part of a nationwide trend'. It is a strong point of this book that while recognizing the peculiarities of the region and their diverse impact, she does not delink them from what's happening in the rest of Maharashtra or the rest of the country. And this is far more important than it might appear to be at first.

She is also clear on how a chain of events deepened the distress in the state. And that the agony of the peasantry was not just due to natural calamity – the generic 'drought' – but also profit- and policy-driven; pursuing a path of 'development' has proved and continues to be extremely destructive.

This is Maharashtra's poorest region, with a per capita income perhaps 60 per cent below the state's average. Also, as she correctly points out: 'Some pockets of these districts (of Marathwada) show some of India's worst levels of nutrition, education, sanitation and healthcare. Child marriage continues to be practised in many pockets of the region.'

Kavitha understands and explains well that Marathwada's drought is deeply connected with issues of inequality. She points to the fact that the difficulties the region's four million farmers face daily are closely linked to its 'historical backwardness and deep-rooted, caste-based exclusions and discrimination'. A place where 'water-intensive industries including beer and paper manufacturing are sited in the region's special economic zones … while lakhs of residents migrate each year', so many of them fleeing acute water scarcity.

She covers the emergence of water markets in Marathwada – and how, if you're a farmer trying to keep his or her fruit trees alive through summer, you must fork out Rs 10,000 for a tanker load of water.

She also sees clearly that this is not just a problem of natural resources and their commercial exploitation, but that water has seriously emerged as an issue of justice. That lends even more substance to her reporting.

There is another important thing this book does. It connects to what is either a non-existent term in official-public discourse, or, at best, a politically correct cliché employed by bureaucrats and ministers: climate change.

As she puts it: 'The impact of climate change is not a faint reality on a distant horizon. It is occurring right now, not far from the big cities but removed enough so that politicians and policymakers in Delhi, Mumbai, Bengaluru, Chennai and Kolkata might miss the urgency, but for residents of Bundelkhand in Uttar Pradesh or Marathwada in Maharashtra, the daily realities now show a clear pattern.

'In these regions there is unprecedented depletion in groundwater levels, long periods of dry weather, a general trend of lower precipitation during the monsoon months or fewer days of rain with long dry spells, continued drawing of groundwater from unsustainable depths, continued drawing of water from reservoirs at record levels of dead storage, repeated cycles of failed monsoons, failed crops, bad debt, desperation and farm suicides. By 2050, India will be a global hotspot for "water insecurity".'

She covers drought as natural calamity but does not foreground that as the main thing. She sees the complex nuances of many interwoven issues. But brings to us the extent of human agency in driving Marathwada's drought. And shows us – in chapter after chapter, instance after instance – the impact on ordinary, everyday human beings of the entire process.

Some policy impacts had nothing to do with drought – but fed deep into every problem farmers and others could possibly face: demonetization, for instance. I was in Osmanabad and Aurangabad three days after the government's grand assault on black money that actually devastated farmers, landless labourers, pensioners, petty traders and many others. The impact on the poor was pretty similar nationwide. In Marathwada and Maharashtra, it came atop a very bad year of drought and scarcity.

There is also the response of those ordinary, everyday human beings – in this case, mostly farmers – across Maharashtra. For the last five years, farmer protests have been growing across Maharashtra, across India. There were major protests in the state in 2017 and then, in 2018, came an event that shook not only Maharashtra but inspired farmers across the country. When 40,000, mostly very poor, Adivasi farmers marched for six days and nights from Nashik to Mumbai, 182 kilometres from their starting point.

Even more startling: for the first time in decades, thousands of middle-class Mumbaikars came out in open support of small and marginal farmers, turning up at the city's historic venue of Azad Maidan, distributing food and water packets. On 12 March 2018, there were doctors setting up a table at which they treated the walkers, many of whom owned no footwear and showed up with bruised and sometimes bloodied feet. Quite a few were also treated for dehydration. Nobody mobilized or arranged for the doctors. They just showed up having watched a few snippets of the march on television. Mostly, of course, from public, non-corporate hospitals. There were young lawyers and law students who showed up demanding they be allowed to file public interest litigation suits against the Devendra Fadnavis government for its treatment of the farmers. There were teachers and students, bank and insurance employees and several other unions, even lots of young techies.

The government, one of whose prominent members had declared that these were not farmers but 'urban Naxals', caved in within hours. They agreed to several of the marchers' demands – obviously only to backtrack on or undermine these issues later.

Many protests followed all over the country demanding that the National Commission on Farmers (aka Swaminathan Commission) report be implemented, especially its recommendations relating to a minimum support price. That report covered an extraordinary number of issues relating to rural and farm distress – including drought and water scarcity. That great Nashik–Mumbai journey inspired over a lakh of farmers from twenty-two states and four union territories to march to Parliament in New Delhi in November 2018.

While Kavitha's focus is, as it ought to be, on Marathwada, she makes important, even crucial, links to those larger events and processes. She shows us how they are connected, how they weave into the region she is reporting on. Her reports are at one level about drought but, at another, even more about loss, about the human condition. This is an important book. Read it.

P. Sainath
Mumbai
December 2020

PREFACE

THE MORINGA STARTED it all, our dowdy old drumsticks, cooked in the fiery dark curry traditional to parts of central Maharashtra. Moringa leaf powder had just about made an appearance in 2014 on the shelves of overpriced produce in Mumbai's organic food stores, and I had made its impersonal, clinical acquaintance around that time as a newbie to health supplements. Then, early in the summer of 2016, I was on a tour of Marathwada, the arid region of central Maharashtra located near the heart of India, when Muktabai Zinjurke of Dhamangaon village invited me home for an impromptu lunch.

As a reporter with the *Indian Express* newspaper, I had been familiar with parts of the region since 2004. I had also begun to visit Beed, Nanded, Aurangabad and Latur districts, all falling within the geographical region known as Marathwada, a couple of times every year since 2011. My unease at this backward part of one of India's most advanced states matched evenly my curiosity about its rather effervescent people. So, when Muktabai served a moringa pod curry that April afternoon when I arrived at her home, dishevelled and dusty after a morning of visiting poor farmers spending the summer months at a 'cattle camp' that offered free cattle feed and water just outside Ashti town in Beed district, it was quite easy to see how her curry got its complex, piquant flavours.

The frail moringa plant supplies a superfood of gigantic proportions. Its leaves and pods are nutrient-packed powerhouses of vitamin C and iron, with anti-fungal and anti-viral properties. But what's more, the moringa is a drought-resilient tree, growing without much fuss in backyard kitchen gardens. And that was the real reason why Muktabai's *shevga* curry sang to me that afternoon a ballad – a Marathi *powada*, perhaps – about that kitchen's geography, climate science, home economics, dryland farming, its inequities in access to water. And about Muktabai's affectionate, soulful cooking.

In a drought-hit region, food is at the intersection of many things. It is in government-mandated relief measures in the form of subsidized foodgrain and free cattle feed, or as midday meals for the children of labourers who have migrated to work in distant towns or farms. It demands that subliminal sustainability choices be made every day in homes, as meals with small carbon footprints.

At the home of Deepak and Kaveri Nagargoje in Beed district's Shirur Kasar taluka, two leafy greens made a lunch in the summer of 2019 particularly memorable: a curry of *chandanbatwa* or *bathua* – *parippukeerai* to Tamilians, goosefoot or *Chenopodium* in English – and *tandalachi bhaji*, another wild-growing leafy green that was cooked with garlic and served still crunchy. Neither plant is cultivated. They grow in thickets beyond farms and pop up as farm weeds sometimes. The Nagargojes, deeply knowledgeable about the region's lifestyle, depended heavily on foraging when, as a young couple in the

1990s, they started a residential school for the children of the district's extremely poor migrant labourers.

Lunch invitations amid a drought are naturally thought-provoking, and these meals crystallized for me the seed of an idea that there are coordinated, integrated tales here that fall just outside the column inches of a newspaper report, stories that might need to be strung together to make sense.

Muktabai Zinjurke was the sarpanch of Dhamangaon village in the extremely drought-prone Ashti taluka. Her home, and those of millions of others in a region battered by cyclical drought, chronic crop destruction and price fluctuations of farm goods, offers insights into how an impoverished society fights to survive.

Food in these homes raises difficult questions. How much groundwater does that spoonful of sugar in a teacup consume? Millets or rice? How much water does either crop consume? Locally grown produce or packaged, imported pulses? But, equally, their lunch-time stories are a collective pushback against state policy that has resolutely refused to consider decentralized planning and local solutions for sustainability, about big dams that submerged thousands of acres of farmland but whose canals remained incomplete, about incidents of farmer suicides in which families couldn't avail the promised financial assistance from the state because, incredibly, the suicide victim had sought medical assistance for his poor mental health.

The evening before I lunched at Muktabai's, I visited Chandrabhaga Khomne, the widow of Kalyan Khomne, in a hamlet called Khomnewadi in Nandurghat village, Kaij taluka, Beed. I was meeting her for the second time, almost a year since her husband had hanged himself from a neem tree on his tiny plot of farmland. She was still grieving.

I returned to Khomnewadi many times in subsequent years, feeling more and more dispirited every time as she began to confide in me. Illiterate, alone and with two sons who were still to find work, she was struggling to even enrol herself for the state's pension scheme for widows. Soon after Kalyan's death, she had decided to grow some jowar, or sorghum, the only sensible choice for her unirrigated farmland, a crop that Marathwada farmers love to describe as one that will survive with even a couple of good showers during the season. But from getting the land transferred to her name and planning how to use the ex-gratia assistance from the state government to insuring her crop without the land title, everything was a challenge. Kalyan's suicide had set her back years, and I saw that I had failed to adequately capture her struggle set in this chronically distressed region despite returning to it every year. Her personal struggles and her family's financial difficulties deepened while I finished a first draft of this book, and in June 2020 one of her sons killed himself, frustrated at the continuing sense of hopelessness.

Farmer suicides in Marathwada had caught up with the scourge in Vidarbha in 2014, but I had not even scratched the surface of this story, having reported the large, hollow numbers without truly understanding what was eating away at the farmers' will to live, what was eroding their confidence as fathers and husbands, and whether professional help and early interventions in facing their inner demons could have stopped them before they reached for the pesticide can or length of rope. In these 'killing fields', as the region's farmland had begun to be called, there was anguish and anger, but also fortitude and kindness.

My city-bred mind, sluggish and lumbering, began to slowly understand that these were not newly minted tales of suffering. The euphoria of the nineties' economic revolution had simply skipped large parts of the country and their people, my new friends among them. Rights-based policy design appeared to get a brief head start but failed to make a dent in Marathwada's backwardness. These districts fare poorly on almost all social indicators of human development – infant mortality, women's literacy, malnutrition. As microfinance and non-banking finance institutions arrived amid a slow destruction of a once-flourishing cooperative banking network, the debt trap became ever more visible – and ominous.

In Chaklamba village of Beed's Georai taluka, Krishna Khedkar, a small farmer with only a school education, captured in a single sentence the months-long struggle that he and his friends had undertaken to demand from local government authorities a sub-canal from Jayakwadi Dam to irrigate their farmland. There were lush cane fields located barely a hundred kilometres from his village, and these were irrigated by canals from Jayakwadi while his own cluster of villages remained at the mercy of the rains. Recounting why groups of men in Chaklamba participated in hunger strikes over and over, Khedkar told me there were too many hungry people waiting for lunch to be served and only one pot of *khichdi*. There just isn't enough dam water to irrigate all of Marathwada, so the scarce resources simply had to be shared more equitably. Chaklamba's struggle for a sub-canal was a demand for a reallocation of resources.

This train of thought chugs right through Marathwada and beyond, to the rest of India where rivers and dams have become sites of serious disputes regarding access, and to various parts of the world where acute dry spells are being reported more frequently than ever. As always, in conversations with farmers, families and farmer-leaders in Marathwada, there appeared to be as many solutions, excellent ones, as there were dilemmas. And nobody to hear either. With these nagging thoughts and half-formed ideas, I kept returning.

Visiting the region more frequently, pressing on with my laboured Marathi that wins me many, many kind-hearted people's indulgence, I began to see a congruence in vastly different people's lines of vision. Historians, water-sector experts, bureaucrats, politicians, trade unionists, farmers and labourers told me the same thing in different ways, that present-day challenges in Marathwada's rural economy are an assimilation of decades' worth of history of inequity, climate change, scratchy policymaking, caste realities and a new globalized world order that failed to take everyone along.

There is an interconnectedness in the scores of news reports from the Indian countryside: a gossamer web that holds together disparate accounts of desolation. These reports often reinforce and give more meaning to one another. Water supply 'cuts' in Mumbai or Auckland, scarcities in Bulawayo or Chennai, shortages in the Godavari in Marathwada or the Amu Darya in Central Asia, flash floods in Western Maharashtra or East Africa – diverse as their people and problems are, a single current of impending tragedy also runs through their narratives. A wide-ranging story of Marathwada's landscapes of loss has unfolded for me slowly over the past decade of travelling there, one that has almost dictated itself to me. This is the story of the people of this land. In that sense, I am no more than a vessel.

Kavitha Iyer
Mumbai
December 2020

ONE

INTRODUCTION

~

'Fish and tortoises know by instinct when the year will be rainy and when there will be a drought.'

—Jataka Tales, 300 BC–400 BCE

TO THE WEARY traveller on a summer day, the town of Jalkot in Maharashtra's Latur district can seem a little like the edge of the world. National Highway 222 melts into dust and blinding sunlight as it rolls towards the neighbouring state of Telangana in south India. Jalkot is little more than a marketplace, a bank, a post office and a bus stop, all huddled around this highway crossing. And from the marketplace, the hot blur of the horizon gives it the appearance of tottering at the precipice of a boiling abyss.

Deep inside Jalkot taluka, on a white-hot day in May 2019, a resident of Umardara village explains what it means to be a farmer in Marathwada. The interior of the gram panchayat office is mostly dark, heavy wooden windows bolted shut against the blistering afternoon. Through a crack in the door, the 44 degrees Celsius midday heat fights its way in slowly. Little circles of condensation from the chilled mineral water bottles on the sarpanch's steel table vanish as soon as they form. Outside, a bullock steps back further into the shade of a solitary neem tree, its neck bell jangling. But for the animal's disquiet, Umardara, with its sun-bleached roof tiles and sunburnt residents, cannot be more still. Or its people more agitated.

'Today, if you have money then you can have water,' says Ashok Bhagwanrao Gutte, a resident farmer. Some farmers in the region have been shelling out ₹5,000–₹10,000 per water tanker to keep fruit trees alive through the long summer. 'If you don't have money, no amount of hard work and none of the fancy new agricultural techniques can help you. Earlier, if you had a crop, you could get money. That was common logic. That was how the world worked. Today, you must have money in order to have a crop. Even for a miserably low yield that will only cause you further losses, you must first have money. Because no money, no water.'

Jalkot recorded a dip in groundwater levels by more than three metres from the five-year average in the summer of 2019. Almost every one of the 350-odd homes in Umardara has a well, but only five or six still have any water left in them. In April, one resident spent a small fortune to dig a fresh borewell, and dug to a depth of 350 feet in rocky land. He struck an inch of water. Umardara is missing its solitary source of water, the Nandanshivani lake on the border between Jalkot and Nanded district's Kandhar taluka. By early May, the lake is bone-dry.

After all, it has been six months now since the Maharashtra government declared a drought following a very poor 2018 monsoon. In regions such as Jalkot, the 2018 kharif crop brought another round of farm losses on account of severe moisture stress. The kharif is the summer crop, served in states such as

Maharashtra by the south-west monsoon. In Umardara, farmers also lost most of their 2017–18 rabi, or winter, crop to attacks by wild boars. Not a single farmer in the village has had even one solidly profitable year in the past decade. Like in all the other villages in the vicinity, they are deep in debt, desperate to receive state dole including in the form of waivers of agricultural bank loans. There is no industrial sector to turn to, so young, educated men who do not want to till the loss-making fields must relocate hundreds of kilometres away.

To set the context correctly, Jalkot is about 550 kilometres from India's financial capital Mumbai. Maharashtra is on the road to becoming a trillion-dollar economy by 2025, according to official government proclamations. But India's most industrialized state is still predominantly agrarian. The share of agriculture in Maharashtra's total gross state value added (GSVA) has been declining, but 53 per cent of Maharashtra's people depend on this sector for their livelihood. (GSVA is a measure of a region's goods and services in value.) And by the state government's own admission, a combination of small landholdings averaging just 3.3 acres per farmer family, rising input costs and uncertainties in climate and farm goods markets has led to farming becoming unprofitable.

In 2018–19, the agriculture sector in Maharashtra grew at 0.4 per cent, continuing a dismal trend. In five of the previous seven years, the state witnessed depressed growth in the sector and 2018–19 marked its worst dip since 2014, another drought year, when it shrank by 16.7 per cent in comparison to the previous year. In the seven years since 2012–13, also a terrible drought year, the sector that employs a large majority of Maharashtra's people posted a sizeable growth just once. This agrarian crisis plays out in homes and on farms by way of continued poor returns for farmers, many suffering deep losses year after year amid repeated cycles of poor, excessive or untimely rainfall until farming is rendered a completely unsustainable livelihood – but 53 per cent of the state has no alternative either.

A Worldwide Trend, Indian Peculiarities

Drought years are now the stuff of nightmares for governments across the world. In the summer of 2020, as the world battled the Covid-19 pandemic, large parts of the world faced another devastation – acute dry conditions leading to impending crop losses, raising questions about food security, food exports, equitable water use and more. Twenty-first-century droughts appear to have outdone historical dry periods in their severity, and the portents are that more horrifying droughts will follow.

In Bulawayo, Zimbabwe's second-largest city, tap water supply was restricted to once a week. The country witnessed a terrible hydrological drought in 2019, and the 2020 rains were not promising. Meanwhile, at the opposite end of the world, residents of Auckland in New Zealand faced stiff fines for using hoses outdoors as the country drew up policies to tackle its worst drought ever.

In Europe, Germany prepared for a third consecutive drought year in April 2020 – the water level in the Rhine was so low that cargo ships' movement was hampered, and experts worried about losses for agriculturists. The German government's Climate Monitoring Report said groundwater had dipped over fifty years, and hot days were on the rise. April 2020 was the third-driest April ever recorded in the country. Other parts of Europe struggled too: crop-growing regions prepared to face persistent droughts, and studies forecast that major crops such as wheat and corn would need to be adapted for dryland cultivation in the future. The spectacular Iguazu Falls, a world heritage site in Argentina, was closed for tourists in May 2020 on account of the Covid-19 pandemic and because South America's drought conditions had left the Iguazu River flow at 13 per cent of average.

In the United States, studies suggested that the continuing drought conditions experienced in various states since 2000 were the mark of a new 'mega-drought', a term for dry periods lasting two decades or more. Scientists belonging to, among others, the Lamont-Doherty Earth Observatory of Columbia University, the NASA Goddard Institute of Space Studies and the Department of Geography at the University of Idaho used hydrological modelling and tree-ring reconstruction to establish that the 2000–18 period was the driest span in US history since the late 1500s. The Missouri, the longest river in the United States, was as dry in recent years as it was during the Dust Bowl period of the 1930s – and more such dry years are foreseen, according to another study. Worldwide, these periods of abnormal dryness put agrarian communities at risk, rendering them most vulnerable to climate shocks and market fluctuations.

In India too, Maharashtra's agrarian crisis was part of a nationwide trend, set squarely and prominently amid a national tragedy in the sector. In 2016, 2017 and 2018, urban India witnessed a new kind of rural unrest, thousands of farmers marching or demonstrating on the streets of major cities. It started in Maharashtra and Madhya Pradesh, both large states with sizeable contributions to India's agricultural production. In March 2018, 40,000 cultivators, almost all of them belonging to indigenous communities, marched 180 kilometres from Nashik to Mumbai. The sea of dusty, sunburnt humans walking peacefully for six days along the highway into the financial capital drew the nation's attention to the downward spiral in the farm sector. Having suffered multiple losses on their farmland, they wanted among other things a complete waiver of agricultural loans, not the conditional waiver announced by the state government, led by the right-wing Bharatiya Janata Party (BJP), in the winter of 2017. Much of these losses were on account of inadequate rains, a particularly horrifying spectre when coupled with other structural problems in the Indian agrarian economy.

In Madhya Pradesh, the summer of 2017 saw five farmers allegedly shot and killed by policemen as thousands demonstrated in Mandsaur in the northern part of the state, seeking government assistance to cope with acute agrarian distress. They were seeking higher support prices from the government purchase machinery for agricultural produce. In the same year, drought-hit farmers from Tamil Nadu travelled to New Delhi, the national capital, and staged a sit-in protest for days at Jantar Mantar, a popular site for demonstrations located not far from Parliament House. Their demands were similar – state aid for an impetus to agriculture and irrigation.

The Swaminathan Commission, or the Commission on Farmers, referred to with veneration at almost every protest march, discussion and television chat show on rural distress, and at every election rally in rural Maharashtra, says in the opening remarks of its fifth and final report submitted in 2006: 'The cost–risk–return structure of farming is becoming adverse to over 80 million farming families operating smallholdings, since the resource-poor families cultivating one to two hectares or less are unable to benefit from the power of scale at either the production or post-harvest phases of farming. Both meteorological and marketing factors influence the well-being of small farm families, who lack the capacity to withstand the shock of either crop failures or uneconomic market prices for their produce.'

Over a decade later, the 2018–19 Economic Survey of India found these conditions worsening. The share of marginal holdings (smaller than one hectare) went from 62.9 per cent in 2000–01 to 68.5 per cent in 2015–16. The area farmed by these small and marginal farmers grew too, from 38.9 per cent in 2000–01 to 47.4 per cent in 2015–16. That means nearly half of India's farmland is tilled by subsistence-level farmers who see the Union government's 'DFI Committee', a new acronym for a team working to 'double farmers' income', as a cruel joke. Indian agriculture saw gross value added (GVA) grow from the negative 0.2 per cent in 2014–15 to 6.3 per cent in 2016–17, and then slow again to 2.9 per cent in 2018–19.

It is against this backdrop that the 'Long March' to Mumbai, as it came to be known, found resonance among agriculturists across India. The march itself followed a swirl of agrarian movements across Maharashtra, fuelled by that adverse cost–risk–return pattern over several years. In March 2016, nearly 40,000 farmers blocked the town square of Nashik city, the townhead of a major agricultural goods hub, seeking to draw the administration's attention to the deepening farm crisis. In October 2016, over 15,000 tribals from Palghar district attempted to gherao the home of a minister.

Then, in November 2016, Maharashtra's farmers, still recovering from the 2014 drought and the depressed crop season of 2015, were again hit when the Union government's move to demonetize currency notes of ₹500 and ₹1,000 devastated the almost entirely cash-run agricultural markets. As the federal bank struggled to introduce new currency, banks in rural and remote areas had no cash for weeks. Farmers were settling for lower and lower rates for their produce if traders offered cash. For most, the next dinner depended on getting ready cash. Traders' cheques couldn't be encashed immediately because banks didn't have currency anyway. Burdened under two crop seasons' losses, small and marginal farmers, who account for the large bulk of Maharashtra's rural population, began to despair.

Six months later, in June 2017, Mumbai, Pune and other big cities experienced acute shortages of farm produce as farmers went on strike, an unprecedented move by the completely unorganized agrarian communities in India. Various groups of farmer leaders got together to convince Maharashtra's farmers to strike, not harvest fresh vegetables, send nothing to the markets, refuse to sell stocks to traders, cancel planned deliveries of farm goods and not even visit their farms. The strike lasted a little over a week. Their demands were the same, the plaintive faces and voices painfully familiar now. The strike ended with the state hastily announcing a loan waiver scheme that would cost the exchequer ₹38,000 crore, with an upper limit and qualifying terms.

The Story of an Indian Drought

If there was one region that awaited the Maharashtra government's loan waiver most desperately, it was Marathwada, where people were known until recent years as being hardy rural folk, made so by a conjunction of factors such as geography, history and sociology. Marathwada, the name given to the eight districts of Aurangabad, Jalna, Beed, Latur, Osmanabad, Parbhani, Hingoli and Nanded, was delivered the worst blow during Maharashtra's four drought years since 2012. In the summer of 2014, when the state government declared 23,802 villages drought-hit, every single one of Marathwada's 8,100-odd villages was among them.

Over three quarters of the 40 lakh farmers of Marathwada own small and marginal holdings, which is no more than four acres of land. Tens of thousands are landless labourers, a legacy of the region's historical backwardness and deep-rooted, caste-based exclusions and discriminations. So, on the one hand, water-intensive industries including beer and paper manufacturing are sited in the region's special economic zones, and on the other, lakhs of residents migrate every year, leaving behind arid and unyielding lands to work as subsistence-level daily wagers in the lush cane fields of water-rich Western Maharashtra, one of the state's administrative divisions.

Every year, thousands of farmer families fleeing water scarcity and near-total loss of crops in Marathwada also arrive in the big cities. Even the squalor and disease of Mumbai's slums hold some promise for them. Migration cleaves families, as children are often left behind with relatives. Families that stay back in the villages walk miles for water in the summer, and duel the lack of livelihood and the depletion of resources including livestock. According to the National Sample Survey Organisation (NSSO), over 50 per cent of agricultural families have outstanding loans from banks and usurious private lenders. This cycle of crop loss and bad credit repeats itself over and over.

But the region was once prosperous. Formerly part of the Deccan, invaded for its incredible riches in the fourteenth century by princes from the Delhi Sultanate, it shows evidence of a once busy trade route around Ter near present-day Osmanabad, on Marathwada's border with Western Maharashtra. Cultural hubs here date back to the medieval era. Relics of the culturally rich dynasties of the south still dot the region – the magnificent rock-cut Ajanta caves built under the rule of the Vakatakas, the Pitalkhora caves during the Satavanahas' reign and other cave monuments that belong to the era of the Rashtrakutas and Chalukayas. Paithan, the taluka where the region's biggest dam is located, was an important cultural hub and home to renowned saints. Aurangabad, formerly called Devgiri and renamed Daulatabad (or Dowlatabad as the colonialists called it), was a crucial administrative landmark too right through the era of the Nizams, the rulers of the southern state of Hyderabad, and then the British.

Lying to the east of the Western Ghats, Marathwada's geography has always challenged farmers. The terrain is mostly tableland and broad plains, with a few hilly stretches in Beed, located at the western edge of Marathwada, and in Nanded, located at the eastern edge, bordering the southern state of Telangana. The plains are alluvial only where they run alongside the Godavari or its tributaries, especially where sediment washed down by the rivers is deposited. Otherwise, the land is rocky, scattered all around the igneous Deccan Trap, a vast pile of basaltic lava that cooled millions of years ago. For farmers in this region, farmland is often thin soil. By necessity, Marathwada's farmers are a skilled lot, having traditionally eked out a crop in a season marked by only two or three spells of rain.

The Godavari hurtles down from Nashik on the north-western edge of Aurangabad district and runs through Aurangabad, Beed, Parbhani and Nanded before entering Telangana. But not even India's second-longest river could alter the historical aridity of the region. There are paddy fields only along the Godavari's banks. While the region historically cropped sorghum and pearl millet, popular cash crops now include cotton and soya bean and, increasingly, sugarcane. Aridity is a long-term climate feature, not to be confused with drought. In Marathwada, the peculiar combination of aridity and cyclical drought, or abnormally dry spells long enough to lead to hydrological imbalances and powerful enough to affect agricultural output, has made farmers particularly vulnerable.

A History of Backwardness

In 2013, former chief of the Finance Commission Dr Vijay Kelkar submitted to the state government a landmark report on regional imbalances in Maharashtra and the backlog of development works in the Vidarbha and Marathwada regions.

It included detailed recommendations for a roadmap to set right the imbalance. His report found per capita income in Marathwada to be at best 60 per cent of the per capita income in the rest of the state. Inequalities within Marathwada may have risen in recent years too, it said. Marathwada fared the worst on the ratio of actual irrigation versus irrigation potential created by dam projects. Only 38 per cent of irrigation potential created in Marathwada is converted into actual irrigation systems with canals and lift-irrigation schemes, compared to 47 per cent in Vidarbha, the region that lies further east. This is despite the traditionally low monsoon precipitation. So Marathwada, which accounts for 31 per cent of the state's area under cropping, uses only 14 per cent of the state's surface water. In comparison, Vidarbha uses 30 per cent of the state's cultivated land and 28 per cent of surface water.

Manufacturing in Aurangabad's special economic zones, with their dedicated hubs for automobiles, pharmaceuticals and renewable energy, has dipped in recent years, but these units are said to have attracted investments totalling $150 million since 2012. And yet, five of the eight districts in Marathwada are categorized as having low relative human development status according to a 2011 state government report. Some pockets of these districts show India's worst levels of nutrition, education, sanitation and healthcare. Child marriage is still practised in many of these areas. The same report said rural poverty, seen through the matrix of consumption deprivation, was highest in these districts. On the urban front too, the incidence of poverty was seen to be the highest here.

The backwardness cannot be blamed on poor political representation – some of Maharashtra's biggest mass leaders belong to Marathwada. Former chief minister Vilasrao Deshmukh represented Latur, former deputy chief minister Gopinath Munde represented Beed. Ashok Chavan, another former chief minister who is also currently a Cabinet member, represents Nanded. But without a deeper people's movement that negotiates with political brokering, such as the cooperative movement in Western Maharashtra, identity-based people's movements for self-realization in these districts were fractured and short-lived.

In the winter of 2019, as a new and unexpected coalition of a far-right political party along with two left-of-centre parties assumed power in Maharashtra, among the very first items up for discussion was another proposed farm loan waiver. A full waiver, as demanded, would cost the state ₹1 lakh crore, equivalent to the annual budget of small nation states. Climate havoc has left almost two-thirds of Maharashtra's farmland wrecked, and approximately one crore farmers in the state have been plunged into another cruel round of farm losses. There is a new edge to farm distress now, as standing crop, tantalizingly close to a year of profits, experienced large-scale destruction when torrential rain poured down just before the harvest. According to the state government's estimates in November 2019, untimely rainfall decimated crops on 88.74 lakh hectares of farmland. That is 88,000 square kilometres, as much as the entire landmass of Portugal or Austria.

Unusual Climate Patterns, Regional Variabilities

Apart from multiple drought years, Maharashtra has witnessed repeated cycles of untimely rains, sometimes hailstorms, since 2012. After a poor 2018 monsoon, as early as the month of October the Maharashtra government declared a hydrological drought in 151 talukas, a little less than half of the state's area, five months before the onset of summer. As the months progressed, it emerged that the bill for drought relief would be close to ₹8,000 crore for 2018–19, in addition to the ₹38,000 crore loan waiver announced in 2017 that was still being implemented. A somewhat tardy south-west monsoon arrived in Mumbai late in June 2019, but news from Marathwada was still ominous. As late as mid-July, the central and state governments were issuing advisories about what appeared to be a deficient monsoon causing the area under kharif sowing to shrink. Water levels in dams had dipped alarmingly, prompting another government advisory, this one from the Central Water Commission, asking the southern and western states, where reservoir levels were 20 per cent lower than a five-year average, to allocate water only for drinking. The advice was to curtail water allocated for industry and irrigation.

Then suddenly, in August, five Indian states were fighting extensive flooding and destruction, including from landslips in the southern state of Kerala. In Western Maharashtra, about two lakh hectares of sugarcane, one of the region's most lucrative crops, was destroyed in Sangli, Satara, Kolhapur and Pune districts in the region's worst-ever floods. As the Godavari swelled, dam gates were opened along its stretch in areas that had until two months ago faced acute water shortages. Days later, the river Krishna breached its banks, and the then chief minister of Maharashtra, Devendra Fadnavis, had to cut short a political tour to take stock of the floods in the very productive farmlands of Western Maharashtra. With rivers furiously in spate, rising dam waters set alarm bells ringing. At least twice, calls had to be made to the government of the neighbouring state of Karnataka to request them to release waters from the Almatti Dam, the first dam over the Krishna on their side of the state border, to give respite to the flooded areas in the Krishna basin upstream.

Even then, farmers in some talukas of Marathwada, located less than 300 kilometres away from the flooded areas, including in Beed, Jalna and Latur districts, continued to wait for better showers. Their kharif crop had begun to show signs of wilting. Meanwhile, Mumbai had enjoyed its wettest July in over a hundred years.

At the end of August, though the monsoon had covered all parts of the country, some regions of Marathwada continued to face water shortages. More than 42,000 animals were still at state-funded fodder camps, mostly in Beed and Osmanabad districts, where farmers had taken their livestock at the height of the summer to access free cattle feed and water. The end of August marks an important farm festival in Marathwada called Pola, a day to dress up farm animals, mainly bullocks, and worship them. With thousands of farm animals still at the fodder camps, several villages skipped the festival entirely, for water to bathe the animal would be difficult to arrange. A few dozen farmers marched to the office of the district collector in Osmanabad, demanding that fodder camps that had been shut down in early August after a few showers be reopened. As the end of the four-month monsoon approached, more than a thousand villages in Marathwada, or about an eighth of the region, continued to require drinking water supply by tankers, with wells and seasonal lakes and streams still dry. It had been a long, arduous summer with no end in sight for these villages.

Then in October, there was a revival of the monsoon as its time came to a close and, in the last week, just around the festival of Diwali, a two-week-long downpour came as an angry reminder of the elements. Exactly one year since the Maharashtra government declared drought in 151 talukas, including most of Marathwada, torrential rains damaged standing crop on nearly half of Maharashtra's farmland. Almost the entire state was affected: varying levels of damage were reported from 325 talukas out of the total 358. Marathwada's eight districts, having borne the brunt of acute water scarcity through the first seven months of 2019, were now hit hard by the floods. On average, about 75 per cent of the region's standing crop was destroyed. Many villages that had begun sowing seeds for an early rabi season found they would have to repeat that investment. Soya bean and cotton, important cash crops, were affected by up to 80 per cent. Some talukas experienced 100 per cent damage to the standing crop.

Cyclical Drought in the Heart of India

Marathwada is witnessing cyclical drought, a phenomenon where acute moisture stress damages crop yield severely every year or every second year for a sustained period. For farmers in these eight districts, extreme climate incidents, such as the November 2019 rains, are only one part of an unfolding narrative of a complex, multifaceted agrarian distress suffered over long durations, impacting daily life, business, deprivation levels, health status and social dynamics.

If there is one narrative that binds the experience of different corners of the country over the past couple of years, it is this: the impact of climate change is not a faint reality on a distant horizon. It is occurring right now, not far from the big cities but removed enough so that politicians and policymakers in Delhi, Mumbai, Bengaluru, Chennai and Kolkata might miss the urgency, but for residents of Bundelkhand in the northern state of Uttar Pradesh or Marathwada in Maharashtra, the daily realities now show a clear pattern.

In these regions there is unprecedented depletion in groundwater levels, long periods of dry weather, a general trend of lower precipitation during the monsoon months or fewer days of rain with long dry spells, continued drawing of groundwater from unsustainable depths and continued drawing of water from reservoirs at record levels of dead storage, or levels so low that water cannot be drained out through spillways but has to be pumped out. There are repeated cycles of failed monsoons, failed crops, bad debt, desperation and farmer suicides. By 2050, India will be a global hotspot for 'water insecurity'.

The severity of hydro-meteorological events should take nobody by surprise. Certainly not policymakers and governments, who are squarely addressed in the assessment reports published by the Intergovenmental Panel on Climate Change (IPCC), a United Nations body. These reports warn that the waning of the world's climate system is 'unequivocal', and the changes observed in recent years 'unprecedented' over millennia. These are not vague warnings of warmer oceans and melting ice. The Fifth Assessment Report said the thirty years from 1983 to 2012 comprised the warmest three-decade period ever in the last 1,400 years in the northern hemisphere. Among a series of specific warnings, it said the impact would be the strongest and 'most comprehensive' for natural systems.

'In many regions, changing precipitation or melting snow and ice are altering hydrological systems, affecting water resources in terms of quantity and quality,' it said, predicting extreme precipitation over most of the mid-latitude landmasses and over wet tropical regions that will very likely become more intense and more frequent. It also predicted reduced renewable surface water and groundwater resources in most dry subtropical regions.

They may not describe or categorize such events as climate costs, but at least ten state governments in India have in recent years written off farm loans. All manner of other bills to the exchequer will have to some day be rechristened as climate costs, ranging from thousands of crores of rupees being spent on bringing water from long distances to arid farmland to ex-gratia payments for farmers who lose standing crop to hailstorms or unseasonal rains. Regions across the world, from African countries to California in the US to parts of Australia, are grappling with drought and repeated climate shocks. The stories from Marathwada would resonate with them all.

The Tragedy Unfolds in Chapters

The 2018–19 drought in Marathwada was the fourth in the past seven years. The combined reality of truant rains, poor irrigation and a vicious geography means that for farming communities in Marathwada not only does one poor monsoon mean an inevitable crop loss but also that consecutive years of losses have broken the backs of self-reliant residents of a once-rich land. Now, with every failed monsoon, Marathwada's small and marginal farmers inch closer to the edge of the precipice. What is cyclical drought? What is its impact on a people and society? And how does Maharashtra's administration respond to this challenge? Chapter Two deals with these questions and gauges the extent of aid poured into the region since 2012.

The poor monsoon of 2018 left wells and borewells so dry that emergency state assistance in the form of water tankers for drinking water supply and fodder camps for livestock had to be initiated as early as January 2019. Despite its dangers, rainfed agriculture is the norm in Marathwada. Irrigation across India is overdependent on groundwater, with about 19 million dug wells, shallow wells and tube wells or borewells accounting for the bulk of farm irrigation. Maharashtra, even with the country's largest number of dams, is not very different. But for Marathwada, the abysmally low level of canal irrigation, incomplete or decrepit canals and long-pending irrigation schemes have meant an even greater burden on groundwater. As extraction outpaces recharge in drylands areas, the danger of deep aquifers shrinking is very real. Chapter Three deals with groundwater and the irrigation infrastructure backlog in Marathwada.

The vicious cycle of loss and desperate measures has led to hundreds of farmer suicides each year. A mental health crisis is slowly unfolding in the region, alongside alcoholism and substance abuse. Chapter Four studies programmes to tackle rural mental health issues that have been unable to quell the tide of suicides. Sugarcane, a lucrative cash crop, has seen a rapid rise in acreage in Marathwada over the past decade, even as experts have pointed out that it is simply too water-intensive a crop for an arid region. Cane fields, those in Marathwada as well as the much larger cane-growing regions in Western Maharashtra, Karnataka and Tamil Nadu, also offer livelihood to lakhs of landless and marginal farmers in Marathwada. Every year, about eight lakh labourers migrate to these regions for a period of three to six months to work on the harvest. Five lakh of these migrants belong to one Marathwada district – Beed. Theirs is an important narrative about the region's poorest. Chapters Five and Eight deal with this topic.

The cycle of loss inflicted by what the Swaminathan Report called a 'constellation of hardships' is sharpened by caste. More on this in Chapter Six. But caste discrimination and marginalization are central to Marathwada's social milieu, and while this chapter looks at land rights for those who suffered historical injustice, the voices of the marginalized ring through almost all chapters – landless or marginal farmers who migrate from Marathwada's villages to live in big cities' slums, women labourers who endure sexual violence, farmers without the means to stand up to private moneylenders' excesses and lakhs of labourers who undertake an annual migration to work in the cane fields of Western Maharashtra – many of these narratives are also about caste as a lived reality in Marathwada.

Chapter Seven considers the policies for industrial development in the region, and the inexorable tide of migrants from these districts who continue to travel to Mumbai and Pune year after year to earn a living.

Chapter Nine focuses on women, one half of the long-suffering people of Marathwada, who are increasingly shouldering the burden of running families as farm incomes bottom out. Even though they are not considered farmers because the land they till is not legally in their name, women are fighting back with a range of solutions from organic farming of food crops to agrobusinesses, home businesses, self-help groups and small savings groups. Their stories are deeply inspiring.

In Maharashtra, drought mitigation works now proceed like clockwork but, conversely, drought risk-proofing is at best guesswork. The Jalyukta Shivar scheme of the Maharashtra state government was to have drought-proofed thousands of villages. A new plan, the Marathwada Water Grid, looks to replicate the desert nation Israel's engineering solutions to convey water across hundreds of kilometres to regions experiencing scarcity. An assessment of these schemes follows in Chapter Ten.

By November 2019, Ashok Gutte of Jalkot, Latur district, suffered extensive crop damage owing to the unseasonal rains. Through the summer, he had talked about an apartheid-like situation over water: if you have the money, you can buy water tankers that come from dozens of kilometres away to water your orchards. As winter approached, the conversation veered to the next generation of Marathwada's farmers, their children, who they hope will not become farmers – because rain and summer are more unpredictable than ever, Marathwada's land more unyielding, and the future seems to forebode water wars and desert farms.

TWO

DROUGHT AFTER DROUGHT

~

'Perhaps the house is not aware that the Nagpur Pact
applies as much to Marathwada as to Vidarbha.'

—Yashwantrao Chavan, Maharashtra's first chief minister,
addressing the state Legislative Assembly in 1960. The
Nagpur Pact of 1953 made way for the creation of a
Marathi-speaking state from parts of the then Bombay
State, Madhya Pradesh and Hyderabad State.

IT HAS BEEN ten days since Manikrao Dhondge died in
Awalgaon, Parbhani district, in central Maharashtra. His
immediate family as well as some friends from nearby villages
have gathered for the final *kriya*, the last of the post-cremation
rituals, near the Gangakhed bridge. Tradition holds that the
ceremony be performed near a river, a dip in its water announcing
the end of the mourning period. The Gangakhed bridge across
the Godavari is where the busy Parbhani–Gangakhed State
Highway cuts across India's second-longest river. But on this

20

day the Dhondges find that the river is not even a stream. It is April 2019, almost the peak of a long drought that set in after 2018's scant monsoon, the next rains still more than two months away. Having slid down a sandy slope, the family sits cross-legged on the riverbed that is barely moist. Sunlight bounces off puddles of muddy water a little further away. A barber gets to work tonsuring the menfolk. A *vasudev*, an itinerant bard and performer of devotional songs, begins singing a melancholic prayer, whirling slowly, his conical hat of peacock feathers slightly askew. A hundred metres away, the sludgy waters that remain of the Godavari swirl around a woman's knees as she flings an old sari into the water to trap fish. Among the mourners, the conversation shifts towards the river, dry for nine months now.

'There isn't even any water for the living,' shrugs Balaji Jayatpal, a family friend. 'Not having water for the dead may not be such a grave thing.'

Almost everyone in Marathwada is in equal measure precise and philosophical while discussing water. Emily Dickinson said water is taught by thirst, a dictum the people of Marathwada appear to understand well. How much has been stored today, how much the tanker trundling down the highway will provide, how many millimetres the farm will require this week, how many cubic metres of storage have been created over the last year, how much a mineral water bottle costs, how much a large animal drinks each day, how many days before doing the laundry will be unavoidable – men and women in the over 8,000 villages of Marathwada always have precise calculations ready. And underpinning all calculations about water and much else is, of course, how much it rained.

The districts comprising Marathwada receive an annual average of just short of 780 millimetres of rain altogether. Roughly equivalent to about two weeks' worth of rains in Mumbai, 780 millimetres through a season is meagre precipitation for agriculture, more so for cultivation that depends entirely on this rainfall. Since 2012, the seasonal average rainfall received by Marathwada has dipped further to about 680 millimetres, weighed down by four drought years. During this period, hundreds of villages recorded rainfall that is short of even those sparse averages by far. In 2014, for example, Parbhani district's seasonal average was 346 millimetres of rain. The same year, Jalna's Ghanasawangi taluka or tehsil, the subdistrict administrative unit, received an average of 229 millimetres of rain. It was just 32 per cent of the usual average. Bhokardan and Mantha talukas, also in Jalna, recorded averages of 298 and 314 millimetres respectively. In Jalna city, home to a clutch of rich Marwari steel magnates, municipal water supply to households came once every fifteen to eighteen days.

Mumbai and the Konkan, where 100–200 millimetres of rainfall in a single day is a common occurrence during the June–September months, are 400–700 kilometres away, and the people here a whole world removed from the realities of central Maharashtra. In every one of Marathwada's villages, homesteads have a large plastic drum (or two) placed outside through the summer, or at the street corner, along with the neighbours', ever prepared for a government-supplied water tanker. Water from the four-foot-tall drum may be all they get for a few days or a week, for bathing, washing, cooking and drinking, and for livestock and pets. Community taps, hand pumps, dug wells and borewells often run dry.

One mid-morning in March 2016, Mailabi Pathan and her much younger neighbour Mannabi Tamboli of Walwad village in Osmanabad's drought-prone Bhoom taluka discuss how much money they had invested in creating additional water storage for the summer months. Standing outside their homes, they say they have been waiting two days for the water tanker to arrive. Like most others in Walwad, they spent nearly ₹2,000 each on purchasing the bigger blue plastic drums and a stretch of hose to be attached to the faucet of the tanker. They made the investment after the 2014 drought, when every single well in the village ran dry and there was no other source of water save the tankers.

The water tankers operated by the government through private contractors, along with fodder camps where livestock can avail of free cattle feed and water, are the most visible signs of rural distress in Marathwada's eight districts. A key element of the government's drought-relief measures, the tankers are mostly 10,000-litre tanker-trucks manned by a driver and a helper. Each truck is assigned a set of villages. Towards the end of May 2019, more than 6,200 tankers were operating across the state, a record high. Of these, 3,233 were plying in Marathwada's villages. In early June 2016, as the state waited for the south-west monsoon, and as the stock of water available in Marathwada's dams dipped to an alarming 1 per cent, 4,003 water tankers were operationalized for the region. At taluka-level government offices, tales flowed of tanker breakdowns as operators ran out of regular tanker-trucks and began to use all manner of makeshift vehicles that groaned on village roads and routinely suffered technical glitches.

Obviously, the tanker economy is central to water supply in these districts, but over the past few years, as rules have been drafted and monitoring of suppliers tightened, the tanker operators have begun to find it difficult to ply their trade too. And with diesel prices rising, the longer and longer trips to locate a still-flush water source renders their business model somewhat unwieldy, even while the number of villages that would not survive a summer without the tankers grows gradually. In 2019, before an unexpected revival of the monsoon as it withdrew, the water scarcity and the absence of adequate water sources for tankers were unnerving enough that the administration in Marathwada drew up plans to dig more than 4,000 new wells in the region, including in the submergence area of dam structures.

'If there are tankers everywhere, you can be sure that the only ones making any money are government contractors. Everyone else would be struggling to protect their crop from total damage,' say brothers Eknath and Dattatray Yadav of Walwad village in Osmanabad.

In the winter of 2015, after the third poor monsoon since 2012, the Yadavs, who together own six acres of land, sowed jowar, or sorghum. The previous rabi season saw a huge shrinkage in the area under sowing, because soil moisture was too low after the failed 2014 monsoon. The kharif season of 2015 was disappointing too. By February 2016, the brothers know they have sunk further into debt. Their jowar crop has failed to grow to the grain-filling stage, and all they have is the foliage, locally called '*kadba*' and used as animal feed. Devastated as they are, the Yadavs are grateful for the green fodder. They will not have to spend money purchasing it. Often, during the summer following a drought that claims fodder crops among its casualties, owners of livestock and operators of fodder camps complain about having to import fodder from Western Maharashtra or Vidarbha, and for all the nuisance of importing and the loss of weight of the green cattle feed as it dries in transit, sometimes only cane foliage is easily available, not the healthiest option for livestock that is already suffering from poor nutrition.

In drought season, the fodder camps that are sanctioned by the government as a relief measure are a distinctive feature of the landscape of Marathwada, visible from the highways every few kilometres. Each camp is designed like a refugee centre for animals, hundreds of bullocks and buffaloes of various breeds, Jersey cows and the Indian 'desi' cows lined up in rows under somewhat elementary sunroofs made of reams and reams of green cloth. The bottle-green sheds stand out against the bleached ochre earth. One corner of the camp is reserved for large stacks of green fodder, where staffers weigh out correct quantities of the feed to be given to every animal owner. Lakhs of animals are checked into the camps every drought year. At the height of the 2019 summer, more than a third of all farm animals in Beed district were at these camps.

In March 2016, in Beed's parched Ashti taluka, eleven-year-old Ghanshyam Ram Pondhe spent restless days battling boredom at a cattle camp near Dhamangaon village. His final exams were approaching, but he spent the torrid, still summer afternoons watching over their three buffaloes. His father's one-acre farm had yielded nearly nothing that season, and he was away working as a daily wager at a metal forging company.

'Mother is out looking for work,' Ghanshyam said sunnily, hopping on to a bicycle for another round of the cattle camp.

As small farmers look for work as labourers, the responsibility of caring for animals at the cattle camps often falls on women, children and senior citizens. The camps are operated by contractors, sometimes local political party workers, who get a fixed payment from the government per animal per day, with different rates for large and small animals. They are expected to provide a minimum fixed quantity of fresh and dry fodder, and water, at no cost to the cattle owners. They are also expected to maintain daily attendance records of the animals, audited by a government staffer every day.

Facilities differ widely at the camps. Some aspire to set new standards, offering everything from electrical charging points for phones and other devices to occasional free meals and sometimes even performances of devotional songs. Others are the subject of complaints such as corruption or inadequate fodder. There are no toilets, and people have to walk home to their villages that may be two to ten kilometres away or, more commonly, do their business wherever the shrubbery offers a little privacy. There are no living arrangements: those spending nights are free to bring a mattress or a cot. By dusk, women begin to leave their animals in the care of neighbours and head home to cook and clean. Many families are too far from home to make the daily trip, and they stay back for days or weeks at a stretch, adorning their patch of the animal sheds with objects in the likeness of home – a doll hanging from a pole, a cloth cradle in a corner for an infant, children's schoolbooks, aluminium pots and pans stacked around a wood-fired hearth or kerosene stove, an occasional radio with a tinny audio of Kishore Kumar and Mohammed Rafi songs, a framed portrait of a deity. Some bring an additional stretch of cloth that is tied around four poles to fashion a sort of private square space where women can crouch and wash themselves. Shaded spots are rare, as the camps are established on large, open fields, so the April and May afternoons pass in a hot haze of exhaustion.

Often, fights break out at the camps over the quantity or quality of the cattle feed, disease outbreaks among the animals, lack of security or shortage of water. In some camps, animal owners have to queue up to collect the feed and run when the water tanker arrives. At all camps, women have to spend long days without any privacy. Some have to hitch-hike to get home. Conversations at the camps revolve wearily around outstanding loans, government aid, the cost of living, the cost of a daughter's upcoming wedding, losses, interest rates, medical expenses, college fees and life at other cattle camps.

This aspect of the government's efforts to tackle a drought, an essential life-support system as it is for animal owners suffering extreme agrarian distress, is also a visceral experience for farmers, a kind of everyday compendium of humiliations. In the summer of 2019, at one camp in Georai, Beed, a man dressed in a long saffron tunic, saffron turban and a saffron shawl, with an elaborate ash mark on his forehead, walked around offering to bless animals and people, for a small fee. A member of the wandering Nathpanthi community of religious performers, he refused to say where he was from and explained that he belonged to no single place: he was a man without a home. 'What difference does it make? I am here now like you are, all of us just as broken as anybody else.'

The cattle camps do indeed become villagers' homes for months during a prolonged drought, and so reflect the whole range of socio-economic distress factors associated with multiple crop failures and a dysfunctional agrarian economy.

At twenty, Samadhan Jadhav agreed he should be working productively instead of whiling his time playing games on his mobile phone at a fodder camp. Samadhan belongs to Osmanabad's Dongrewadi village, but spent most of April 2019 at a fodder camp in neighbouring Beed's Nandurghat village, about ten kilometres from home. He completed his Class X a few years ago but dropped out because his family could not afford his education after their seven-acre farm was ravaged by multiple crop failures 2012 onwards. The Jadhavs owned three bulls, so there was no milk to sell either.

'I just spend the day alone, waiting for the next day,' he said.

Though milk prices fluctuate, those with milch animals checked in at these camps are doubly grateful for the free fodder. The few litres of milk they sell every morning is usually their sole income at such times. Some camps have the facility of a daily visit by milk traders, and with a profit of a few hundred rupees a day, the stay at the camp is worth the effort. But Sadiq Shaikh, the contractor running one of Nandurghat's three fodder camps, had a different problem. The single borewell on the land, 500 feet deep, was close to running out. He would have to buy water, and investing in water tankers for the camp would throw off his calculations.

~

GROWING UP IN Beed district's Georai, Satish Ghadge Patil, forty-two, had heard stories of his ancestors owning elephants and horses. Now his family is left keeping up appearances as one of Chaklamba village's 'Patils', not just a powerful and upper-caste Maratha community but also formerly the revenue-collecting foot soldiers of the Nizam in this region. He earns a meagre ₹6,000 per month as a temporary employee with state telecommunications company Bharat Sanchar Nigam Ltd. 'The name is all we have left now,' says Satish, a father of two sons.

His father owned 400 acres of farmland a few decades ago, on which they grew sorghum, pearl millet and green chillies, he says, seated in the family's once palatial '*wada*', a traditional home built in an architectural style typical to Maharashtra's moneyed class. Their *wada* was once the cynosure of all eyes in Chaklamba village. The house is no longer the imposing structure it used to be, larger cement and brick buildings dwarfing it now, but its stone-walled exterior, weeds growing from the cracks in places, and the adjoining temple stand testament to his story. The *wada* is now subdivided to make separate living spaces for his late brother's home, his own home and a space rented out to a little shop fitted with a pressured water jet where a man washes the villagers' mucky motorbikes.

One room is also converted into the village's solitary beauty parlour, where Satish's wife Anjali works alone, tweezing young women's eyebrows into shape and applying make-up on brides and bridesmaids. Hers is the only salon in the four or five villages in the vicinity, so she is busy most days, earning ₹200 to ₹300 per day on average. Despite the rampant sniggering in the village at a Patil's wife cutting other people's hair, an occupation traditionally reserved for lower castes, she has kept at it, for the family cannot forego the income. Their farmland stretches out behind this multi-use homestead, all eighteen acres of it, jointly owned by him and his brother's family. At nine acres, in a region where the average size of landholding is less than three acres, Satish is still a medium landowner.

In the mid-1970s, the senior Ghadge Patil was forced to give up hundreds of acres of land under an agricultural land ceiling law enacted in the post-zamindari era of the 1960s. Later, as agriculture became increasingly unsustainable, he parcelled out more land to continue living well until he died in 1984. 'A woman in a Patil household couldn't possibly be seen working, so my mum started selling her belongings, and, finally, she began burning her zari saris to collect the melting silver to be sold,' he recounts. Then came year after year of farm losses, until he finally took up the low-paying job. 'What were we, and what have we become? This bothers me day and night,' Satish says. His sons, twenty-one and eighteen, do not want to be farmers. 'First the law, then our own troubles, and now the climate – nothing has gone our way.'

Near identical hardships and a depletion of assets alongside successive farm losses have affected farmers across class and community. A mid-level Nationalist Congress Party (NCP) politician living near Undri village in Beed's Kaij taluka said that the thirty days between 15 July and 15 August in 2014 had witnessed four farmer suicides in the Nandurghat village nearby. 'Two of them were big farmers,' he rued. He himself had to wind up an enterprise after running into heavy losses. A few years ago, he had purchased three dozen buffaloes from Rohtak in the northern state of Haryana, famous for its dairy businesses and livestock market. But sourcing cattle feed in Beed proved to be an expensive proposition; it was either difficult to get or was too expensive. As the costs of medicines, fodder, other nutrients and the wages of the men from Bihar he had hired for the animals' upkeep added up, he ended up with ₹25 lakh in the red – and his children's planned foreign education in peril.

Cyclical drought spares nobody and no part of a village economy. As the seasons of crop failure add up, sometimes without break, the landed class experiences impoverishment as a slow burn, over decades. Small and marginal farmers, who constitute the bulk of Marathwada's agrarian community, face indigence rapidly, sometimes losing land or other farm assets to private moneylenders as institutional credit dries up. Among the poorest or the landless, many have no alternative but to relocate to big cities where they can engage in unskilled labour. Their narratives are evidence of the impact of cyclical drought and distress, but they also provide, in a wider sense, proof that policy measures by successive governments to tackle the agrarian crisis have never succeeded in empowering the rural community to be better prepared for the next disaster.

The crop failures themselves are indicative of the deepening distress. Damage to crops was reported in the years 2008, 2009 and 2010 too, mostly on account of excess or unseasonal rains, pest attacks and, specific to horticultural crops in 2010, excess heat. Together, according to data from the Maharashtra government, ₹1,023 crore was disbursed for climate-related crop damage in those three years. This aid for farmers who suffer losses on account of drought or floods or other causes does not cover their anticipated revenue or profits. It is calculated as an average of losses per hectare, usually with an upper limit of two hectares per farmer, with variable factors computed such as what crop was damaged, whether the land was irrigated, and whether the lost crop was a horticultural crop.

In 2011, a long dry spell during the monsoon, coinciding with the crucial growth phase of the kharif crops, led to damage to the cotton, soya bean and grain crops. Aurangabad division, which comprises Marathwada's eight districts, saw the most extensive damages. As ex-gratia assistance, the state government estimated that it would pay out a total of ₹1,585 crore, including the lion's share to farmers in the Aurangabad division. In June 2013, the government assessed further damage to the 2011 kharif crop, the 2011–12 rabi crop and the 2012 kharif crop, all damaged to varying levels owing to inadequate rains. A sum of ₹2,183 crore was disbursed as aid; the largest slices of the aid pie were for Marathwada (₹514 crore) and Vidarbha (₹642 crore). Just eight months later, the damage to the rabi 2012–13 season was assessed and aid worth ₹824 crore announced.

Curiously, while every government resolution announcing aid says, in bold letters sometimes, that no further aid would be forthcoming as damage has been assessed by the district administration and estimates drawn up accordingly, demands from the divisional commissioners of various regions continue to emerge for years afterwards. Officials say this is on account of fresh evidence of damage that occurred after initial assessments and poor data gathering at the village end of the chain. So, for instance, for damage incurred on soya bean, foodgrains and cotton fields in 2011, additional aid requests continued to pour in and were accepted until the end of 2013. Similarly, for damage to the rabi crop of 2012–13, requests to the state's revenue department continued until March 2015.

The year 2014 saw government assistance paid to farmers for crop damage record a sharp increase. It was one of the worst drought years in recent decades, as farms suffered extensive damage and complete crop loss in large regions. In all, 23,802 villages were declared drought-affected, and all crops had sustained damages. A detailed matrix of crop-wise and region-wise payments was prepared, and the bill added up to ₹4,803 crore. The largest payments would be, once again, to Marathwada's farmers, totalling ₹2,032 crore. That was followed by the assistance package for crop damage incurred during the 2015 kharif season, in which 15,747 villages were affected. The total assistance to be paid was estimated at ₹3,578 crore, including ₹2,069 crore to Marathwada. Once again, various divisions including Aurangabad continued to ask for additional assistance until at least January 2017. The year 2016 was a good one, and 2017 was average though a few talukas were declared drought-hit in the rabi season. The next severe drought year was 2018, when Marathwada and North Maharashtra were the worst-hit regions. Total assistance paid was estimated to be about ₹8,000 crore. In addition, assessments were also simultaneously underway for damage to standing crop caused in August 2019 and October–November 2019 by floods and untimely rains.

So, between 2009–10 and 2015–16, Maharashtra's farmers needed assistance totalling well over ₹11,300 crore. It only works out to a few thousand rupees per farmer. And yet, these aid packages along with crop insurance pay-outs have become a sort of lifeline for small and marginal farmers who also find it difficult to get work as farm labourers during drought years. This repeated pattern of crop loss and ex-gratia payments translates into a heavy dependence on state dole, a vicious cycle in which state aid is a key element of household finances, and further investments in the farm for the next season are uncertain, depending on whether institutional credit is available and whether the risk of taking money from private moneylenders at usurious rates becomes an inevitability. Farmers in regions that have experienced repeated cycles of total crop loss feel a sense of helplessness as the last vestiges of self-esteem and pride are slowly lost.

The frequent drought years have lent a degree of smoothness and immediacy to the way the drought relief manual kicks in and government measures that focus on amelioration are rolled out rapidly. Apart from the water tankers and the cattle camps, the measures may include all or some of the following: waivers of land revenue, waivers of electricity bills for agricultural pumps and other farm equipment, waivers of exam fees for Class X and Class XII students in the affected regions who are taking the state board exams. These measures have, without a doubt, deepened and widened in their scale and reach, while expenses on them have grown. The expenses on the 2012, 2014, 2015 and 2018 fodder camps and water tankers alone were a couple of thousand crore rupees.

The relief measures and the expenditure on them are watched and interpreted carefully by political leaders, the ruling party and the opposition viewing ground realities with astonishingly diverse lenses. The former chief minister of Maharashtra Ashok Chavan, who belongs to Nanded in Marathwada, spoke extensively about the drought during the Congress party's Jan Sangharsh Yatra, or People's Struggles Tour, in 2018, a year before state assembly elections. When Chavan set out on the third leg of the tour, in the last quarter of the year, just after the monsoon, Marathwada was facing acute water scarcity. Water levels in Jayakwadi Dam had dipped to 37 per cent, live storage in the Majalgaon Dam, a critical water source for Latur city, was down to 2 per cent.

'The government has failed to release water into Jayakwadi Dam. Now, they have even started diverting water from the Upper Penganga project, leading to potential water scarcities in Nanded, Yavatmal, Parbhani and Hingoli,' said Chavan. He had already met the state's chief minister and governor. 'We will not allow this government to destroy the future of Marathwada,' he declared, adding, 'Chief Minister Devendra Fadnavis's flagship Jalyukt Shivar is the biggest scam in Maharashtra. Crores have been spent, but there has been no change in the water table. Jalyukt Shivar has only benefited private contractors affiliated to the BJP.'

In October 2020, following an audit report by the Comptroller and Auditor General of India (CAG) raising queries regarding the quality, cost and impact of work undertaken through this scheme, the new state government decided to launch an enquiry into the Jalyukt Shivar Abhiyan.

~

PEOPLE IN MARATHWADA often compare recent droughts with the agricultural drought that ravaged the region in 1972. The most common refrain is that in 1972 there was no foodgrain, but water scarcity was not as cataclysmic as it is today. In his detailed report on the 1972 drought, senior bureaucrat S.E. Sukthankar, then secretary for urban development and public health, and chairman of the 1972 Fact-Finding Committee for the Survey of Scarcity Areas of Maharashtra, wrote that the state was no stranger to drought and scarcity. After all, not only is ancient literature replete with instances and descriptions of famines but historians over several centuries also took note of calamitous years, either on account of war-ravaged farmland or epidemic or hostile weather. Sukthankar's report said there had been fourteen widespread famines in India between 1765 and 1858, and in twenty of the forty-eight years between 1860 and

1908, scarcity conditions or drought prevailed in one region or the other.

Maharashtra was not spared, though district and state boundaries altered a few times since. Some of the pre-eighteenth century famines had been christened rather imaginatively too. The 1460 famine of the pre-Shivaji era, for example, is known commonly as the Damaji Pant famine, named after a revenue officer who, according to legend, was so pained upon seeing the condition of the poor and the hungry that he distributed foodgrain from the royal granaries of the rulers of Bijapur. When he was asked to pay up, the deity Vithoba is said to have manifested in the Bijapur durbar. The year 1630, in which the boy who would become Maratha warrior king Chhatrapati Shivaji was born, coincided with the 1620–31 drought that devastated large parts of the Deccan, Gujarat, Khandesh (parts of present-day Northern Maharashtra and adjoining Madhya Pradesh), Berar (present-day Amravati in Vidarbha) and Daulatabad (present-day Aurangabad). Parts of the Deccan continued to experience famines and droughts, scarcities on account of floods and markedly long dry spells.

Until the 1970s and '80s, the problem of scarcity usually referred to grain because goods in those days did not have the unobstructed flow across district and state borders that they enjoy today, globalized markets were nascent and water was nowhere near as precious a commodity as it is now. Drought, then, meant scarcity of food and commodities and, consequently, depressed trade, poor availability of work and sequential impacts on the rural economy.

'During the years 1970–71 and 1971–72, scarcity conditions of a severity and magnitude not witnessed before in this state in the last fifty years or more exhibited themselves,' said the government resolution of 1972 through which the fact-finding committee was appointed. Forty-five years later, the scale of destruction and the cost of relief are much greater, but the narratives of a long period of scarcity – and the recommendations on 'the potentiality of measures for the permanent rehabilitation' of affected areas – remain starkly unchanged. In other words, the recommendations of 1972 are not significantly different from present-day discussions on amelioration and prevention.

According to data provided by the Maharashtra government, 208 villages were affected by scarcity in 1960–61, growing to 16,155 in 1965–66 and 24,797 in 1972–73. In 2014 and 2018, the number stood once again at more than 20,000 villages. The characteristics of drought-hit villages in 1972 were not very different either. The report notes that there had been a failed monsoon, withering of crops, reduction in water supply, rise in prices of grains, increased demand for employment, loss of cattle, fodder shortage and migration. Relocation of large numbers of people for employment and survival had been common in the region, at least since the nineteenth century, as it remains to date. For example, between 1876 and 1878, according to the report, one lakh people left Poona for Bombay city, some travelling to Gujarat. Many also left Poona for Berar, then a separate province. There are two aspects of the 1970s' scarcities that Maharashtra has clearly outgrown. These are starvation deaths and epidemics. But the one that is new to the current decade's crisis, and finds no mention in 1972's government documents, is, of course, farmer suicides.

On drought relief works, the state's response goes back to 1880 when the Famine Commission asked provinces to draft their own famine codes. The Bombay Presidency wrote up its manual on how to tackle famines in 1885, a rulebook that stayed in force until it was replaced in 1954 by the Draft Scarcity Manual. Making a distinction between 'famine' and 'scarcity', the Famine Code categorized employment generation works to be taken up under each – local, municipal and village level works if they were affected by scarcity and larger works on road building and irrigation projects for people affected by famine, with separate pay scales for each. Under the latter, 3,000 or more would find work at each of the large construction sites and would have employment for longer durations of time.

The distinction was deleted in the 1954 Scarcity Manual, and the Sukthankar Committee explains why. 'Prior to Independence, however, the emphasis continued to be on the disagreeable works in order to keep the labour attendance to the minimum and thus effect economy in expenditure on relief works,' the report said. Other measures it assessed were the fodder camps and the method of calculation for payment of ex-gratia relief to farmers who suffered crop damage. In all, nothing dramatically dissimilar was in operation during the 2014 and 2018 droughts, barring the now ubiquitous water tankers.

This is despite recommendations by multiple panels in preceding decades that the focus must shift to prevention, better water budgeting in agriculture and better preparedness. Through the nineteenth century, governments the world over considered famines and scarcity conditions to be calamitous events that could not be prevented, and assumed that administrative responses would be limited to easing people's suffering through providing employment and relief in land revenue taxation. Some assumed economic growth and development, alongside long periods of peace and the end of war combat that destroyed farmland, would automatically cause scarcities to wane. This may be acceptable wisdom today in part, but only because the nature of scarcity has transformed. What Marathwada now faces too often is a region-wide acute shortage of water after very long and cyclical dry spells, rapid depletion of groundwater resources alongside lower precipitation, and the inability of governments to either identify long-term measures to cope with the changing circumstances or locate alternative methods of providing water for dryland agriculture.

Further, there are key recommendations made in 1973 and earlier that remain pertinent. Sukthankar in 1973 recommended industries for scarcity-affected regions to provide employment, but those with low water requirement. Among Marathwada's biggest industries today are beer and paper manufacturers, both water-intensive. More than a decade before the Sukthankar report, a fact-finding committee set up in 1958 under bureaucrat N.S. Pardasani to study scarcity-affected areas of the then Bombay State and to suggest long-term ameliorative measures submitted its report in 1960. It said, 'The policy of adopting scarcity measures when scarcity is already in sight has to be substituted by a strategy of development based on appraisal of basic deficiencies in the economy of an area.' Pardasani could have written that today.

Skewing the equation further for drought-hit farmers is the inability to access fresh institutional credit owing to a combination of factors – old outstanding loans, the erosion of the formerly robust network of cooperative banks, and banks' reluctance to lend to farmers amid fluctuations in farm goods' prices and climate, which render them credit-unworthy. In the winter of 2017, the Maharashtra government announced a ₹38,000 crore agricultural loan waiver scheme, in response to demands from farmer organizations and political parties, to provide relief to lakhs of farmers reeling from the effects of cyclical drought and multiple farm losses over the previous three to five years. Despite this waiver, weeks before sowing operations began for the 2019 kharif season, farm leaders were still complaining that small farmers were unable to access fresh loans. The government summoned bank representatives and urged them to meet crop loan targets set for them. Crop loans are a 'priority lending' sector, but credit uptake in this sector was low in 2017 and 2018 too.

Only two all-India loan waiver schemes have been implemented to date – the Agriculture and Rural Debt Relief Scheme (ARDRS) of 1990 and the Agricultural Debt Waiver and Debt Relief Scheme (ADWDRS) of 2008. While the ARDRS had a spending cap of ₹10,000 per farmer, the ADWDRS did not have an upper limit and instead based eligibility criteria on the size of landholdings, providing higher benefits to small and marginal farmers who received a full waiver of the eligible outstanding sums. Since 2014–15, however, at least ten states have announced loan waivers, and Maharashtra's package is among the largest, fuelled not just by political considerations but also repeated drought years and a collapse of agricultural produce prices alongside consistently growing agricultural input costs.

The loan waivers, by themselves at least, have failed to energize the rural economy in the absence of wider policy corrections. Simultaneously, fresh credit uptake has remained depressed, not least because the small and marginal farmers are unable to get loans the subsequent year, even though one of the fundamental objectives of loan waiver schemes is to open fresh loans for farmers. The RBI has said loan waivers may be effective only if credit allocation to beneficiary households is not impacted, but credit flow to agriculture has, in fact, historically slowed down after a loan waiver scheme is implemented. This could be because beneficiary farmers limit themselves to rather conservative farm operations in the face of continuing economic distress, thus demanding less credit. But equally, a clean slate for fresh credit could nudge them towards intensive, ambitious operations to cover previous losses.

There is no data to say which is more true, or less false. But there are scores of accounts of farmers and farmer leaders who say, year after year, that poor farmers face difficulties in getting fresh loan applications processed. There is also the supposed 'moral hazard' ruling finance companies' credit decisions – borrowers preferring to default in anticipation of a loan waiver and banks thus preferring to reallocate lending to lower-risk borrowers.

The pre-sowing weeks are when farmers purchase seeds, fertilizers or other farm inputs. Some may install drip irrigation systems, purchase or hire a tractor or a bullock, build a storage shed or dig a fresh well. In 2018–19, banks met only 54 per cent of their crop loan targets. Even as bankers claimed the demand for credit was low, the state raised the crop loan target for 2019–20 by about 2.5 per cent from the previous year's unachieved targets.

According to data from the state government, disbursement of loans in the agriculture sector by banks in 2016–17 was ₹96,906 crore, a year-on-year growth of 33 per cent. In 2017–18, loan disbursement fell to ₹48,857 crore, a negative year-on-year growth rate of minus 50 per cent. In 2018–19, loan disbursement totalled ₹67,914 crore or 79 per cent of the targeted ₹85,464 crore, a year-on-year growth of 39 per cent. Of these disbursals, crop loans in 2016–17 were ₹42,173 crore, a 3 per cent growth over the previous year. In 2017–18, crop loans totalled ₹25,322 crore, a 40 per cent dip from the previous year. In 2018–19, while the target for crop loans was ₹58,324 crore, actual crop loans disbursed totalled ₹31,327 crore, 54 per cent of the target but a 23 per cent rise since the previous year. Officials said that achieving 54 per cent of the target in 2018–19, from 47 per cent in 2017–18, was still good news. But despite credit camps for farmers and the push by the state, 2019 credit uptake remained poor until July–August, even as farmer leaders said banks' claims of a dip in demand for agricultural loans were deliberately misleading.

In March 2020, the Maharashtra government asked banks to extend fresh credit to farmers who were eligible for the previous loan waiver scheme of the state government even if their loan accounts were still to be credited with the write-off sum. But banks appeared recalcitrant, and until the end of June, when sowing for the kharif season was well underway, only about 30 per cent of crop loan targets had been disbursed.

In Marathwada, where in addition to debilitating losses in the drought years of 2012, 2014, 2015 and 2018, there were also losses on account of February–March hailstorms, untimely rains, dry spells during the crucial growth phase of crops and at least a couple of skipped rabi seasons since 2012, the effect of continuing dependence on state aid, shrinking access to institutional credit and depressed price realizations for farm goods has been a continuing and deepening distress.

A drought is defined simply as an extended dry period when there is inadequate or no rain. As a cyclical phenomenon, returning several times over a span of a few years, and combined with traditional aridity and low average precipitation, its impact on Marathwada has been crushing. When this cycle started in 2012, nobody really noticed, for 2011 had not been a great year either. The drought crept upon a region where poor rainfall is hardly a matter of note. In the years since then, its effect on the region's economy, its people and their lifestyles, migration, social relations, mental well-being and physical health has been quick and brutal. Eight years after the 2012 drought, the people of Marathwada have begun to ask pertinent questions. What is the cause of cyclical drought? Could the causes be anthropogenic? Could agricultural drought and especially long periods of poor or no cropping further contribute to higher temperatures through evapotranspiration? As intense heat waves become an annual feature, the Aurangabad divisional administration has had to issue citizen advisories in recent years. But to the people's questions, policymakers find themselves short of answers.

T H R E E

THE EVER DEEPER SEARCH FOR WATER

~

'शत वरुषें अनावृष्टि । तेणें जळेल हे सष्टि'

(There will be a hundred-year drought. All creation will burn.)

—From *Dasbodh*, a seventeenth-century spiritual text by
Saint Samarth Ramdas, who was born in
Jalna, Marathwada

THE SIXTH DAY of the spring month of Phalgun according to the Marathi-Hindu calendar is Nath Shashti. Nath is for Sant Eknath, a sixteenth-century poet and one of Maharashtra's most prominent scholar–saints, and Shashti the sixth day of the lunar cycle. Nath Shashti, marking the day when the saint consigned himself to the waters of the Godavari more than four hundred years ago, is one of the state's biggest religious gatherings. On Nath Shashti every year, a sea of devotees floods the banks of the Godavari River in the town of Paithan, the saint's birthplace. Paithan is located a little south of Aurangabad, the Marathwada region's biggest city, and is also the location of the region's biggest dam, Jayakwadi.

The pilgrims who gather on Nath Shashti belong mostly to the Warkari community of devotees and arrive in *dindis*, large groups of men and women singing prayers as they walk hundreds of kilometres to their destination. In every *dindi*, a few women carry on their heads a 'Tulsi Vrindavan', a square clay planter with a potted sapling of the Tulsi or holy basil plant. The men are in dusty whites, a Gandhi cap or brightly coloured turban on their heads, cracked and callused feet in slippers. The women are in bright cotton saris. Some are barefoot, some walk a few dozen kilometres a day in the scorching sun while observing a dawn-to-dusk fast. Many carry brass *manjira*, a pair of mini-cymbals.

Sunburnt and relentless, the Warkaris are Maharashtra's most visible practitioners of Hinduism's bhakti tradition, belonging to various caste groups, bonded by their devotion. On Nath Shashti, about five to eight lakh Warkaris arrive in Paithan and settle down in the state-erected *rahutis* or large tents just off the stepway into the river. Through three days of piety and renditions of *abhangas* or devotional songs, every devotee takes a dip in the Godavari.

In March 2019, on the sixth day of Phalgun, however, the Godavari had run dry. Before the Jayakwadi Dam authorities released just enough water for the holy dip, a few thousand devotees had gathered at the reservoir site, to send up collective prayers to the river. Between *abhangas* composed by Sant Eknath and Sant Bhanudas, they had sad accounts to share with one another, of wells and lakes and ponds running dry wherever they walked. One group from Jalna's Jafrabad taluka stopped en route near Rohilagad, a large village with a historical fortification nearby. Thousands of mosambi, or sweet lime, trees were being hacked down in Rohilagad. There were bright yellow rig trucks rolling down the highways to drill borewells, many with number plates from Andhra Pradesh or Telangana. There were farmers among the pilgrims who couldn't sow a rabi crop in the winter of 2018 because their farm wells ran dry. Drinking water tankers made their way into the soft bed of dam structures to refill. It was a blow-by-blow recap of acute water scarcity that the Warkaris saw unfolding before their eyes along the journey.

Rohilagad is indeed a tormented village, one of many in Jalna's Ambad taluka, once one of India's largest hubs for sweet lime production. Along the gently undulating Aurangabad–Rohilagad–Ambad road in February 2019, there are acres and acres of ghost trees on both sides, blackened skeletons with no foliage, some with shrivelled, hard brown fruit. Many are being hacked down with chainsaws. The payment to the operator-owner of the saw is ₹100 per tree that is cut. Some of the larger farmers own over 1,000 fruit trees each, so it's cheaper to use farm labour. By some estimates, through the drought years of 2012, 2014, 2015 and 2018, at least one lakh sweet lime trees were hacked down in Ambad and nearby areas. About 15 per cent of Rohilagad's farmers own sweet lime orchards, and every single one of them has run into deep losses trying to keep their trees alive to fruit in another season. Most of Ambad's orchards depend on wells and borewells for irrigation, and the long dry months of 2018 have bled the horticulturists dry.

Babasaheb Khadelbharad owns 2,500 trees of sweet lime and pomegranate. Through January and February 2019, he purchased fifteen tankers of water every day, at a cost of a little over ₹9,000 per day. 'I'm supposed to be one of the bigger farmers here, because I own large orchards. But I had to take a loan of ₹2.5 lakh from a private moneylender just to buy water for the orchards. And there's no guarantee that the trees will survive until the rains come. Water could just run out, and along with it my luck and my last paisa,' he says. He was among the first farmers in Rohilagad to experiment with pomegranate and has been a horticulturist for decades, almost since the early 1990s.

Marathwada's traditionally low-value agriculture, comprising mostly cereal crops, was transformed in the early 1990s by, among other things, the advent of fruit orchards, enthusiastically promoted by the government through a horticulture development programme that was linked to the state's employment guarantee scheme. Though a department of horticulture was functional even in the 1980s, it was with the 1990–91 government scheme that horticulturists really arrived on the scene as significant contributors to Maharashtra's rural economy and to local employment of farm labour, also creating jobs in packaging and transportation, and so on. Lakhs of small and marginal farmers took up horticulture too, for the financial gains were rather obvious. Not only were the revenues many times those of traditional crops, but the scheme also sweetened the deal. Farmers with up to four hectares of land could grow pre-specified fruits with a large subsidy from the government on the initial investment and on fertilizer and labourer wages for the first three years. Thousands of orchards sprang up in the next few years, all heavily dependent on groundwater for irrigation.

Khadelbharad was among that first generation of fruit-growers. His friend Vilas Takle, who owns 550 sweet lime trees, remembers Rohilagad's horticulturists transporting water from a well twenty kilometres away in the year 2004 when their own wells dried up. But in 2019, Takle doesn't have the money to irrigate his orchard, and he's preparing to cut down a few hundred trees on his thirty acres.

Rohilagad's other big crop is cotton, and the 2018 kharif season ended with a severely stunted cotton crop in the region, standing no more than eighteen inches tall. Those who earlier harvested fifteen quintals from an acre found they only managed to pick three quintals in 2018.

Everyone is shocked at the village wells all drying up in tandem. In nearby Sukhapuri, Annasaheb Bhakad hacked down half of his 2,000 mature fruit trees during the 2012 drought. About twenty kilometres away, in Math Pimpalgaon, a much smaller farmer, Devidas Jige, also a teacher and an active leader of the Communist Party of India, cut 150 sweet lime trees from among his 300 in 2012. He also had to dip into his savings to dig a new well. The little additional water, and cutting down half his orchard, helped keep the remaining half alive. All along the Ambad belt, Jige says, the dipping groundwater levels are a grave challenge for horticulturists who are facing the prospect of their orchards shrinking every year.

In 2017, 2018 and 2019, the Maharashtra government's Groundwater Survey and Development Agency (GSDA) recorded a dip in the Static Water Level (SWL) of all eight districts in Marathwada. SWL refers to the normal water level in a well when it has not been pumped for water for several hours. In 2019, the dip in SWL was by 1.84 metres from a five-year average. The most alarming depletion was in Osmanabad district, where there was an average dip by 3.21 metres; followed by Beed with a 2.59-metre dip; and Jalna, where an average 2.42-metre dip was recorded (all five-year averages). In addition, officials say the pace at which this depletion is occurring has picked up in recent years.

In the summer of 2019, the GSDA found that 279 of the state's 353 talukas recorded depletion in groundwater levels as compared to a five-year mean, and Marathwada was the worst affected. Of seventy-six talukas in Marathwada's eight districts, seventy-two recorded serious depletion in groundwater levels, with more than 5,000 villages of this region recording a dip of more than one metre from the five-year average. In 1,467 Marathwada villages, mostly located in Beed, Osmanabad and Aurangabad districts, the depletion recorded was by more than three metres from the five-year average. Six months earlier, the GSDA's post-monsoon 2018 survey showed 4,385 villages in Marathwada witnessing a water table dip by over a metre, while the April 2019 survey set that number at 5,145.

A 2016 study of Maharashtra by the Central Ground Water Board showed that when mean groundwater level for the period May 2005 to May 2014 was compared with the groundwater level in May 2015, only 11 per cent of test wells showed a rise in water levels in the pre-monsoon period, while 89 per cent showed a depletion of groundwater. Of the test wells where a decline was observed, 20 per cent wells showed a significant fall in depth, by more than four metres. This decline in water level was observed most prominently in Aurangabad, Jalna, Nanded, Parbhani and Hingoli districts, all in Marathwada.

The comparison of the 2015 post-monsoon groundwater levels with the decadal post-monsoon mean was equally disturbing. The mean groundwater level for the period November 2005 to November 2014 compared with the groundwater level in November 2015 showed that 58 per cent of test wells recorded a decline. Again, a significant decline of more than four metres was recorded in Nanded, Latur, Osmanabad, Solapur, Pune, Beed, Ahmednagar, Aurangabad, Jalna, Parbhani, Nashik, Dhule, Jalgaon, Akola, Buldhana, Nagpur and Amravati districts. That includes seven of Marathwada's eight districts. The study made a list of recommendations, particularly regarding groundwater development, increasing water use efficiency, and groundwater regulation in the over-exploited, critical and semi-critical watersheds.

~

WHY IS GROUNDWATER depletion such a concern? According to the Asian Water Development Outlook report by the Asian Development Bank in 2016, nearly 89 per cent of groundwater extracted in India is for the irrigation of farmland. Simultaneously, in Maharashtra, 65 to 70 per cent of all irrigation is from groundwater. This is so because canal irrigation has a rather low potential in large parts of the state, especially in Marathwada, where no more than 25 per cent of farmland is served by canals built to operate on gravity. So, the bulk of water drawn to irrigate crops is from wells and borewells.

And ultimately, all underground water is derived from rainwater – there is simply no other source of water that replenishes the centuries-old aquifers deep in the earth. Logically, therefore, extraction of groundwater can be sustainable only so long as it stops short of the quantum of recharge. In 2018–19, a severe drought year for large parts of the state, every district of Marathwada recorded considerably less rain than its average. For example, Beed district, with the lowest average of the eight districts at 666 millimetres, received an average rainfall of 335 millimetres through the entire season.

Calculating groundwater recharge is a complex process that assesses the nature of the soil, its porosity, the slope and other factors. But back-to-back drought years have generated a series of groundwater assessment reports that show progressive depletion of the groundwater table. A quick recap of average rainfall in the eight districts together, as a percentage of the long period mean, shows Marathwada received 118 per cent in 2009 and 125 per cent in 2010. After that, it crossed 100 per cent only twice, 110 per cent in 2013 and 113 per cent in 2016. These are not heavy rainfall years by any estimation, of course – the average in millimetres was just 854 mm and 879 mm respectively. Then there were two middling monsoons, in 2011 and 2017, when the region received an average of 84 per cent and 86 per cent of its long period average rainfall, or 655 millimetres and 673 millimetres respectively. That leaves four years of acute drought – 2012, 2014, 2015 and 2018. And right through this period, groundwater extraction has grown exponentially, evidenced simply by the numbers of farmers desperately trying to find water by investing in more borewells.

The express warning to government about groundwater depletion came at least as early as 1972, when farmers told a fact-finding committee studying that year's terrible drought that water levels had begun dipping in their dug wells over the previous decade. And this was long before borewells, dug to depths of 300–700 feet, became so popular in Marathwada. Groundwater depletion in Marathwada is particularly treacherous because the Deccan Trap, the ancient volcanic formation around which Marathwada's towns and villages are scattered, is a vast pile of igneous rock, formed through flows of basaltic lava through fissures and volcanoes over millions of years. Much of this geology is not suited to hold moisture. Marathwada's rivers are also unlike those in the Vidarbha region lying further east. Barring the whimsical Godavari that cuts right across the breadth of Marathwada, the other rivers in the region are relatively humble, prone to run dry during harsh summers.

In Aurangabad district, about twenty kilometres from Khuldabad city, sixty-seven-year-old Dattubai Khandagale tended to her jowar crop in a dried-up part of the submergence area of the Girija Medium Irrigation Project, a dam across the river Girija. Dattubai lives near Yesgaon, one of the villages where project-affected farmers were rehabilitated after the dam's construction. But in November 2018, at the start of what was to be a long drought after the failed monsoon that year, farm wells had already yielded only muddy sediment, so Dattubai and a few other families returned to the land that used to be theirs.

This is fertile soil, still moisture-rich on account of a healthy groundwater table around the river, but for dam oustees the reason to return is as much sentiment and faith in their patch of earth as it is the readily available soil moisture. 'If I don't do this, I'll have to skip the rabi season,' Dattubai reasoned. Three wells just outside the submergence area are key sources of water for nearby villages, but those were expected to dry up in the next few months, the next monsoon still seven months away.

Aurangabad holds a special place in Marathwada's groundwater history, for as far back as 1612, the just-founded city was equipped with an underground aqueduct that would be named the *neher* system, stone and masonry conduits that ran underground and along the surface of plains. Designed by the city's original founder, Abyssinian mercenary-turned-military-leader Malik Ambar, the *neher* network was further developed by later rulers. According to historians, the system was built upon discovery of an underground water source in the region of the Harsul Lake, north of the city in that era, now in Aurangabad's suburbs. More historical dots may be joined to comprehend the seriousness of the water scarcity, and particularly the significance of groundwater, dating back to the pre-Nizam era in the Deccan.

Legend has it that in the mid-1300s, Delhi's ruler and the Tughlaq dynasty founder Muhammad bin Tughlaq decided that the Delhi Sultanate must have a second capital and declared it would be Deogiri, later named Daulatabad, literally translated as the home of wealth, comprising the hill and valley adjoining Khuldabad. To mark his seriousness about developing Deogiri as a second Delhi, he dispatched a series of intellectuals, artists and administrators to the Deccan, wise men who would be his advisors in the south. Eventually, this cultural migration into the

Deccan saw a large posse of Sufi saints make Khuldabad their home. Dotting present-day Khuldabad are the tombs of those who brought Sufism to Marathwada, whitewashed domes and arches glinting prettily against the sun. But historians also record a long drought stretching from 1396 to 1408, a twelve-year dry spell called the Durgadevi famine. As wells and rivers dried up, the Deccan witnessed large-scale migration during those years as people moved north and north-west to the Malwa region and Gujarat. During the Durgadevi famine, entire districts south of the Narmada were depopulated as rulers made efforts to resettle people in kinder geographies.

'We have had droughts in Marathwada since the time of the Nizam,' says Sanjay Vithalrao Adhane, in his mid-thirties, a large farmer in Khuldabad taluka's Vikramgaon village, where he owns 110 acres. 'But even during very bad rain years between 1990 and 2010, we never lost both crop seasons in a year. The decision to not sow a rabi crop was always very, very rare, reserved for the worst circumstances when the rains failed and the wells were entirely dry too. In recent years, though, we seem to skip rabi crops every third or fourth year. There is just no moisture in the ground.' In 2017, Sanjay harvested 1,700 quintals of corn; the same farmland yielded 150 quintals of corn in the 2018 kharif season. Sanjay incurred losses worth ₹7 lakh and chose not to make any investment in the land through the winter crop season, a luxury that small and marginal farmers do not have. He usually sows wheat during the rabi season. 'In previous years, we would just dig a new well. Now it makes no sense to do that.'

But left with no choice by September 2019, when it appeared that the south-west monsoon had skirted around Marathwada again, the administration drew up plans to sink nearly 4,000 new wells in the region, including many in dried-up water bodies such as dam submergence areas and river beds. A post-monsoon deluge in October and November 2019 brought respite to the immediate water crisis, but in September 2019 the region still had an overall rain deficit of 19 per cent. Work on surveying locations for the new wells began and drinking water tankers continued to draw water from the dead storage of reservoirs, levels so low that water must be pumped out. In Osmanabad, seven of eight talukas had recorded less than 400 millimetres of average rainfall by the end of September; Beed and parts of Latur fared only a little better.

But farmers who are digging fresh wells are not striking any luck. In Georai taluka of Beed, the fodder camp at Chaklamba, where over 500 animals lived for nearly four months in the 2019 summer to avail government-funded cattle feed and water, was set up on farmer Sanjay Shitole's land. The Shitoles had scooped out the bottom of their savings and had taken loans to raise ₹3 lakh to purchase water for their sweet lime orchard, at ₹5,000 to ₹10,000 per tanker. 'We planted 450 trees on a two-acre plot in 2007,' says Hirabai Shitole, Sanjay's soft-spoken mother. Her elder son died a few years ago after an unexplained cardiac arrest, and she now mothers her three grandchildren, all studying in nearby towns. 'But there is no income,' says her husband Trimbak Shitole, wearing a hearing aid and looking older than his sixty-five years. Sweet lime trees take five years to begin fruiting, and just about then the years of drought set in. They also had cotton and bajra on their twelve acres for the 2019 kharif season, both severely stunted at harvest time. The Shitoles drilled four borewells over the past three to four years, and struck no water at all, even after drilling to depths of 250 to 300 feet. Their sweet limes trees were all but charred by the time the first showers arrived.

These poor rainfall years and the austerities that surround them are like a loathsome habit in these parts now, difficult to shake off. Drinking water is supplied by water tankers to villages where wells have dried up and where there is no other proximate source of water available. These tankers come from farther and farther away every drought year. The number of tankers operationalized during a drought year has grown manifold. Where 365 villages and 177 hamlets of the region needed tankers during the summer months of 2010, the numbers rose to 2,972 villages and 1,017 hamlets in Marathwada during the summer of 2016 and then to 2,502 villages and 848 hamlets in May 2019. So abiding are these appurtenances that many residents remember the last seven to ten years as one long drought with a couple of small breaks.

~

THE HORTICULTURIST KHADELBHARAD and marginal farmer Dattubai have a vague sense that there has been irreversible damage done to the water that, for decades, kept springing in their wells. Khadelbharad set up orchards based on a calculation of how much water his land can provide. Dattubai reserves as her last hope her former farmland that is now in the submergence area of a dam. Sanjay Adhane skipped a rabi season to curtail his losses. These and lakhs of other Marathwada farmers have a shared experience of their chief source of irrigation giving up mid-season. Owing to their heavy dependence on groundwater, when a well dries up, their entire world is imperilled.

Apart from a bevy of government directives and at least one order passed by the Maharashtra Water Resources Regulatory Authority (MWRRA), the apex body that decides on water allocation, tariffs and more, the Maharashtra government in 2009 passed the Maharashtra Groundwater (Development and Management) Act, a law that was to facilitate and ensure sustainable, equitable and adequate supply of groundwater to various categories of users. It was meant to be a law protecting public drinking water sources. Through a state groundwater authority and district-level bodies, it was to manage and regulate the exploitation of groundwater. The law appears to leave nothing to chance, making provisions for gazette notifications of areas where groundwater has improved or depleted, prohibiting any effluent release that could contaminate groundwater, funding the recharging of groundwater in suitable areas, and prohibiting the drilling of deep-wells.

Nothing could be farther from the ground reality. Across Marathwada, tens of thousands of borewells have been drilled since the enactment of the law. These are commonly up to 400–500 feet deep, a far cry from a half-hearted 2015 order by the MWRRA that prohibits borewells deeper than sixty metres (197 feet) in notified areas where groundwater depletion is serious. There are also detailed rules regarding how many borewells may be permitted per square kilometre. Nobody really cares about the rules. In fact, as droughts become more frequent and amid the intricate web of distress for farmers in these districts, the rules may be entirely unenforceable as farmers cope with rising losses and adopt more despairing means. In any case, the detailed rulebook that will enable implementation of the law was never drafted.

'The rules under the Groundwater Act are yet to be formulated and therefore there is no implementation,' says Pradeep Purandare, water policy expert and former associate professor at Water and Land Management Institute, Aurangabad. Once implemented, these rules would mandate registration of existing wells and permissions before fresh well-digging, itself requiring a huge awareness campaign among the lakhs of farmers who own wells and borewells.

Meanwhile, neither the law nor any state survey attempts to analyse the depleting groundwater levels in the region with changing cropping pattern, the growth of high-yield crops, cash crops, horticultural crops and sugarcane. At least a few new agricultural growth strategies may have to share some of the blame. The 2013 Kelkar Committee Report on regional development imbalances in the state pointed out that with the increased use of groundwater, Marathwada has been able to improve its gross cropped area index. The question may now have to be asked about whether cane cultivation and the boom in horticultural cropping are important anthropogenic climate change factors to take note of. Also, even more vexing for the farmers, if groundwater levels are on a permanent downward trajectory, what happens to the orchard and plantation owners who raised their families out of poverty by taking up the government's 1990–91 horticulture development scheme? Will they sink back into insolvency or subsistence farming?

Instead of wrestling with these questions, policy has ended up being contradictory. For example, a hugely popular subsidy scheme to build 'farm ponds' or artificial water storage structures on farmland has seen lakhs of farmers pump water out of borewells to fill into the ponds, in a sort of race to store groundwater in impervious plastic-lined ponds before the wells dry up for the summer.

As the 'groundwater authority' of the state, MWRRA is expected to notify and de-notify watersheds or aquifer areas, prohibit the drilling of deep-wells as well as withdrawal of groundwater from the existing deep-wells, ensure registration of well owners, make registration of drilling rig owners and operators mandatory, and so on. Following a 2016 amendment to the MWRRA Act, the post for a Member (Groundwater) was created for the purpose of implementing various functions as envisaged by the Maharashtra Groundwater Act, but with the rules not notified, there is little the MWRRA can do to enforce the law.

Almost a decade ago, a World Bank report calling for pragmatic measures to address the problem said Maharashtra is one among seven states that account for more than 80 per cent of critical and overexploited groundwater blocks in India. The World Bank is a partner in the Union government's proposed Atal Bhujal Yojana for more sustainable groundwater management. According to the bank's 2010 report, about 15 per cent of India's food production is currently dependent on groundwater use that it described as 'unsustainable', the situation most precarious in rain-fed or drought-prone areas where subsistence farming is prevalent. Overall, up to a quarter of India's harvest is estimated to be at risk due to groundwater depletion, the report said.

India is the world's largest user of groundwater, estimated in 2016 by the Asian Development Bank to be 251 cubic kilometres per year, more than twice the extraction by the US or China, or a quarter of the global total groundwater use. Marathwada may well be at a pivotal moment in its groundwater history. By 2050, the world population will balloon to 10 billion. As the world's hunger grows, so will groundwater extraction, and tremendous water insecurity. Since the 1990s, this is precisely what Marathwada has already witnessed, agriculture developing on the strength of groundwater extraction, now unsustainably so, with apparently no safeguards to prevent a catastrophe.

THE AGONIES OF A HARDY PEOPLE

~

'मेला नाही अजून आभाळ, आत्ताच आशा सोडू नका'
(The sky is still alive, and so is hope.)

—Lyricist Balaji Madaningle for singer Sandeep Bhure's
album of farmers' songs, 2018

SUICIDE. THAT'S HOW daddy died, hanging from a tree, says Kajal*, eleven years old, her saucer-like eyes the colour of charcoal, her gaze steady, dry-eyed. She didn't know the details then, but over the last four years she's gleaned most of it from her mother. A resident of Sonimoha village in Beed's Dharur taluka, her father had received repeated visits from one of the village's private moneylenders. Marginal farmers with less than an acre of land, they were left in penury by the drought of 2014. That year there was no crop, and no work available on the fields.

---

* The second names of some people have been withheld in this
chapter to protect their identity

When the creditor threatened to kill his wife and children, Kajal's father felt he had run out of options.

'Sometimes I also feel tense. If there is no rain, how can we grow a crop and repay people we've taken money from?' Kajal asks. Her maternal uncle also died by suicide a year later. Kajal, who wants to study medicine someday, is shy, slow to make friends. Occasionally, she confides in her mother about a sense of foreboding. On those days, she says nervously, she feels that perhaps her father and uncle had been left with no choice.

Two hundred kilometres away, in Donwada village of Aurangabad, Mahadev is resolute that he won't work to earn a living. Addicted to cannabis, he works when he needs to buy a hit, earning enough to board a state transport bus to nearby Phulambri town to meet his drug dealer, before another slow ride home in the stuffy bus with bottle-green rexine seats. In his late thirties, the brooding man has lived out the worst already. When his father died suddenly five years ago, Mahadev, thrown into the role of chief provider for the family, began to be more and more dependent on a little daily delirium. As his addiction grew, he began to suffer bouts of psychosis, says his mother. Mahadev became something of a terror in Donwada, increasingly unpredictable and unstable, growling menacingly at children sometimes and at other times completely catatonic. Twice he had to be hospitalized, admitted once to Government Medical College, Aurangabad, and the second time, when he began to turn violent, to Yerwada Mental Hospital, Pune. His condition was serious enough that the hospital insisted on an affidavit from the family declaring he was being left there at their behest.

Over a year since, Mahadev is somewhat stable, on medications, but moody. He will not work. The family's land yielded nothing in 2018–19, and his wife Jaya – his second wife after the first reportedly left him when his addiction rendered him unable to work or earn – works as a labourer to provide for the family, which includes his mother and their toddler. Their hardships are compounded by the fact that Mahadev is intensely suspicious and jealous when she chats with male labourers she's working with.

'There are fights every day at home; it's stressful,' Jaya says brightly, sticking the comb into her sari petticoat's drawstring to braid her glossy black hair. She is slim, vivacious and incredibly cheerful despite her unenviable life. Jaya understands that her husband may have a mental illness, though she's not sure what exactly can be done to ease her own pressures of being the family's sole provider.

Distressed as Mahadev is since 2013 when his father died right after a severe drought year, he has never tried to take his life. But Shobha Pawar, the ASHA worker in Donwada, once doused herself with kerosene and threatened to strike a match. The ASHA, or Accredited Social Health Activists, commonly referred to as 'ASHA workers', are the last-mile providers of public health services to women and children in villages, ensuring that nutrition, medical care and other health-related government schemes for mothers, infants and children are properly administered. Shobha's husband was drinking almost every evening, and the arguments at home, along with the stress of her own work as a caregiver, seemed altogether insurmountable one afternoon almost four years back. Something snapped that day and Shobha took a jerry can of kerosene, marched to the porch and, crying, emptied it on herself, she recollects. She bounced back within minutes. 'I recognized that I was under tremendous stress and I should watch myself.' She told friends sheepishly that she had briefly lost control over her actions, but she didn't seek any medical help for it. It didn't seem necessary, she says.

Donwada village has not recorded any farmer suicides in a while, but Shobha reads and re-reads several times the questionnaire filled by farmers under the Prerana Prakalp, a state government project to address farmer suicides through early diagnosis of mental health problems. Neither the questionnaire nor the project's training material says anything about the normalization of severe stress in villages hit by repeated years of drought and total crop loss. The truth is that everyone is on edge, Shobha says. Amid the grinding expenses that continue without a decent farm profit in years, the questionnaire is supposed to help her identify those suffering from extreme stress and those who may have contemplated self-harm. But pretty much everyone is anxious around here, Shobha reasons, and the women much more so.

Donwada is one of a handful of villages in the Aurangabad region from where men, faced with deepening losses on their farms, turned into cotton traders selling in Surat, Gujarat. Shobha says the men's long absences every year when they are away in the cotton markets outside the state, their reported encounters with sex workers during these journeys coupled with the everyday worries over finances and children render Donwada's women particularly vulnerable to stress. 'The men are away for several weeks at a stretch; sometimes they make repeated trips. In addition, they're drinking, even more heavily if the times are bad – such as a bad monsoon or a poor crop or low crop prices.' The women, as a result, are increasingly vulnerable to anxiety and a sense of hopelessness, especially those facing socio-economic stressors compounded by domestic bitterness. 'I talk to the women, I tell them to be like Durga, the goddess,' Shobha says. 'But that one time, I myself couldn't deal with the weight of it all.'

Until 2012, farmer suicides were reported most commonly from the Vidarbha region at the far east of Maharashtra. Since the drought of 2012, however, Marathwada has accounted for a large percentage of the state's farmer suicides too, the first indication of an unfolding mental health crisis. For the public health department, formulating policy to address it will depend on accurately assessing the extent of the problem, but going by Shobha's narrative, the suicides alone are not at all an adequate indicator of how deep the region's traumas run. As years of losses pile up, Marathwada's farmers face other problems alongside the socio-economic impediments. Young men are unable to find brides willing to live in drought-hit and water-scarce villages. Farmers with daughters, land disputes, an ailing loved one, those beginning to enjoy a couple of drinks every day or struggling with addictions – the stressors may be dissimilar but they all may be in need of help. At the same time, seeking treatment for mental illness bears a stigma. Also, access to mental healthcare remains poor. These two factors reinforce each other, building a stoic workforce with slowly eroding mental health that continues to take pride in enduring silently. Until they no longer can.

On the policy side, meanwhile, the absence of any state intervention through the 2005–15 decade when agrarian distress deepened rapidly means that there is no evidence or data of any manner regarding the mental health and well-being of 53 per cent of the state's population that continues to depend on agriculture for their livelihood.

Half of that 53 per cent, the women, are even more reluctant to approach medical professionals for coping with stress or anxiety, though they may need the help urgently and though there is some awareness generated about access to such healthcare. In Osmanabad's Kalamb taluka, for example, when non-governmental organization Paryay along with ActionAid India conducted a survey a couple of years ago, they found as many as 765 women in just thirty villages who were either widowed or abandoned, almost all of them in the age group of twenty-one to thirty-five years. Many were farm widows, and many more had lost husbands to fatal road accidents.

'They usually have one or two children, and often they lack support from their own families and from their husbands' families. The marital family routinely holds them responsible for their son's misfortune. There's a saying in Marathi – "*manus mela, jag budhla*", meaning the world sinks when a husband dies. That's because a new widow has no idea at all how she will survive, raise her children, what she will feed them next week. It is most common for her to feel that it would be better if she were dead too,' says Sunanda Kharate, forty-six, single after a brief marriage at the age of eighteen. 'Their mental condition is tremendously delicate. They need someone to hold their hand and show some support.'

Paryay and ActionAid India mobilized these women into self-help groups, roped in another NGO to help build houses for those who had been rendered homeless, and also organized legal aid for those who needed it. According to Kharate and Paryay founder Vishwanath Todkar, the experience of unexpected widowhood or abandonment along with already difficult life circumstances portends for these women a fragility in financial matters and social functioning, especially in the initial year or eighteen months after the trauma. Some problems they are able to talk about, to support groups or friends or, as in this case, an NGO, such as requiring proper shelter or a source of income or discussing a financial matter they have never tackled earlier. But other problems, such as the adverse effects of the transition on their mental health and their many depressive symptoms, often go unrecognized. Many share privately that they experience sleeplessness, mood swings or inexplicable fatigue, but are not able to discuss them with a doctor. Sexual health is not even considered. It's clear from conversations with them that they struggle to comprehend that these symptoms could be evidence of mental health challenges that can be treated.

During a discussion in Rajya Sabha in May 2016 on drought conditions affecting eleven states in the country, Member of Parliament Supriya Sule of the NCP chose to speak about these less obvious outcomes of the continuing drought. 'Latur and Marathwada are on television every day, and I have to say with sadness that the Maharashtra government has failed to control the situation. A twelve-year-old girl Yogita Desai was going to school and collapsed – she died of dehydration. Rajashree Kamble, aged ten, was helping her mother fetch water from the well and fell into the well. There are other instances. We hear often of farmer suicides, but this time there is a rising incidence of children dying in these conditions. Swati Pitale used to take a state transport bus to go to college. She needed ₹260 for the bus pass, but did not receive any assistance despite her attempts. The nineteen-year-old wrote a letter to her father before she took her life, in which she said she didn't want him to also have to incur the prohibitive expenses on her wedding. She thought it was better to die,' Sule said.

Swati's suicide in Latur the previous year had been widely reported, and the Maharashtra State Road Transport Corporation (MSRTC) decided to give free bus rides to all college students in all eight districts of Marathwada. The then transport minister, Diwakar Raote of the Shiv Sena, also announced that the scheme, which was to benefit 4.6 lakh students, would be called 'Swati Abhay Yojana' as a tribute to her. But no government agency has yet begun to study the unique stressors that the region's young women face.

Little Kajal, Shobha, Mahadev and Jaya are all unable to express their stress levels clearly. And their stress appears to them as being unexceptional and commonplace, owing to the increased vulnerability of rural communities suffering from a long-term socio-economic downturn. What's more, unless there are outward manifestations of the problem that affect their behaviour or functioning, these individuals will not consider getting medical help. And though Marathwada's cyclical drought since 2012 and the burgeoning numbers of farmer suicides have finally prompted the government to attempt to systematize help-seeking for mental illnesses, there is no data or surveys on how common depression, stress or panic attacks are in rural communities. Without any information on the extent of the problem, policy interventions are limited to being band-aid assistance. Though multiple government decisions have declared an intent to curb the self-harm of the distressed farmers, efforts to mitigate rural Maharashtra's mental health risks are conspicuously missing.

~

IN SEPTEMBER 2015, after months of deliberation, the Maharashtra government's public health department launched its first-ever project to address the problem of suicides among any class of people, the Prerana Prakalp, designed specifically to reduce the incidence of farmer suicides in the state by providing access to mental healthcare in fourteen districts where the problem was most acute. The genesis of the project lay in a wide-ranging slew of initiatives by the government that July, involving as many as nine different departments that would seek to reassure distressed farmers. Calling it a 'result-oriented new package', the government resolution dated 24 July 2015 said various departments would implement policy measures to 'stop' farmer suicides under the programme 'Krishi Samruddhi', or prosperity in agriculture. The ideas were not exactly path-breaking, but the attempt to gather various departments' efforts together for a single outcome was a first.

The public health department was tasked with appointing mental health experts in hospitals, including in rural hospitals, and ensuring that distressed families were provided counselling and treatment. The cooperatives department would set about restructuring farm loans, the water conservation department would prioritize villages in the severely hit Yavatmal (Vidarbha) and Osmanabad (Marathwada) districts. In addition, the agriculture department would set up community groups and train them in modern agriculture, the water resources department would increase the area under irrigation and the animal husbandry department would promote poultry and aquaculture, among others, all independent schemes to tackle a single problem – farmer suicides.

Another government resolution was issued the same day, announcing special programmes for Yavatmal and Osmanabad districts' farmers. These included setting up committees including village-level teams that would draw up lists of distressed families. Awareness campaigns would be organized in villages with one-act plays highlighting the importance of community wedding ceremonies (as a cost saver) and 'bhajan-kirtan' for those requiring de-addiction help, among various other concentrated measures such as training in low-cost agricultural practices. Osmanabad and Yavatmal each would receive an additional fund of ₹10 crore for the purposes of this resolution. This government resolution also announced the 'Baliraja Chetna Abhiyan', or farmer awakening programme, a focused three-year effort to use theatre, music, devotional songs and literature to inspire hope and positivity among farmers. Together, the two government resolutions were meant to appear impressive.

The public health department envisioned the Prerana Prakalp for the fourteen districts – the eight in Marathwada, along with Akola, Amravati, Buldhana, Washim, Wardha and Yavatmal in Vidarbha – based on the two resolutions. Under this project, a team comprising a psychiatrist, a psychologist, a community nurse, a psychiatric nurse, a psychiatric social worker and a data entry operator were appointed for each of the districts. This team then trained the ASHA workers, primarily women, numbering between 1,000 and 2,500 per district. The ASHA workers were equipped with a readymade questionnaire seeking to assess a person's mental state with pointed questions regarding thoughts of self-harm, listlessness, psychosomatic ailments, etc. These women went door to door in villages over a period of several weeks, getting farmers and family members in every household to fill the questionnaire.

The answers indicated what category the person was to be placed in – no, mild, moderate or severe depression. Depending on the category, the Prerana team's doctors began to meet patients who needed counselling or medication, referred some patients to a primary health centre, a rural hospital or the district hospital where a psychiatrist could assess them. The Prerana team also trained medical officers manning the primary health centres, the famously ill-equipped and poorly staffed village clinics. They began to refer patients showing signs of severe depression for admission to the district hospitals where five beds in the male medicine ward and five in the female medicine ward were designated as makeshift psychiatric health wards. Full-fledged psychiatric care wards are otherwise available only at the state's twenty-four hospitals that are attached to medical colleges.

Over the next four years, social workers and psychiatrists worked slowly and steadily as awareness about *manasik arogya* or mental health grew bit by bit. In 2018 and 2019, the teams were almost always busy, and the flow of farmers complaining of stress, anxiety, panic attacks and somatic reflections of stress grew, as the ASHA workers, paid a small fee to report cases where depression is likely, also began to understand the pattern of reporting and follow-ups.

In Osmanabad's Civil Hospital, the Prerana team is supported by the state's first 'Vyasanmukti Kendra', or de-addiction centre, run under the National Rural Health Mission. Counsellor Harshada Narayankar sometimes gets patients from the hospital itself, farmers sick with a rasping cough may show up at her clinic and declare their readiness to kick the tobacco habit. Or some show interest when she conducts outreach events in crowded places such as bus-stops, with posters and information on de-addiction. She refers some to a nine-month de-addiction programme. 'Reasons for substance abuse are multiple and very common,' she says, 'such as farm loans, losses, peer pressure, depression.'

Osmanabad's clinical psychologist Suhas Shinde of the Prerana team saw eighty-five patients with severe depression during 2018–19. He uses standard testing materials to decide which patients require only out-patient treatment or counselling, and severe cases that require medication and a hospital stay. He tests for personality disorders, suicidal ideation, sleep disturbances and addictions. He instructs some to return for follow-up consultations. If they don't return, he tries to videoconference with them via ASHA workers' smartphones. On Tuesdays and Thursdays, he and his team make visits to rural hospitals in Washi, Kalamb, Umarga, Paranda, Tuljapur, Bhoom and Lohara. The team conducts a monthly 'death audit' of villages and visits families where a suicide was clearly on account of farm distress.

'Mainly we need to check if any other family member is also depressed – we found one such case last month,' Shinde said in the summer of 2019. They also map high-risk areas, normally coinciding with drought-prone zones, and hold counselling camps in villages in these areas.

Prerana Prakalp has been in operation since 2015. According to the state government, at least 2,000 medical officers and 20,000 ASHA workers were trained on how to respond to cases of mental illness in these districts. In three years, these personnel conducted surveys in the fourteen districts, and their hotline, 104, received more than 26,000 calls. At least 8,000 farmers or farm labourers were treated as in-patients at various hospitals. Thousands of farmers were given counselling, and follow-up enquiries by ASHA workers continue in these cases. But in 2018, even as the project continued, there was no dip in the number of farmer suicides being reported every year. Top officials and doctors in the public health administration began to concede as much in private.

The number of patients accessing the Prerana team's care also grew rapidly, indicating, on the one hand, the need to buttress the team strength and, on the other, evidencing the acute need for mental healthcare in Marathwada. The farmers screened in the Prerana teams' out-patient department (OPD) clinics in Marathwada's eight districts grew from 7,31,191 patients in 2015–16 to 41,00,778 in 2017–18, before dipping to 28,38,691 in 2018–19. Those referred for in-patient treatment rose sharply, from 48,095 farmers in 2015–16 to 1,62,234 in 2017–18 and 2,74,369 in 2018–19.

'The scope of the project is wide but there is sadly little impact on the number of farmer suicide cases, which has continued to grow,' says one of the topmost government officials in Maharashtra's health administration, with a long career in the department. 'We didn't review its efficacy, unfortunately, though the Government of Maharashtra funds were availed.' He says he spoke to ASHA workers himself to understand what was going wrong. Asked how they didn't notice something amiss in the families that were subsequently visited by tragedy, they admitted they had no idea at all and had been taken by surprise, indicating that either their interview questionnaire and assessment methodology were falling short, or that they were simply not trained enough to glean from conversations the underlying despair.

The other interesting thing is that the numbers do not show a decline in the one clear 'non-drought' year of 2016. The four-year period from January 2011 to December 2014 saw 6,268 farmer suicides in the state, the number rising steeply to 12,021 in the next four-year period between January 2015 and December 2018, according to data from the state government. In 2018, the highest numbers of cases were in Vidarbha (1,297 in eleven districts) and Marathwada (947 in eight districts). In comparison with incidence rates in other states of India and other countries too, Maharashtra records relatively more farmers killing themselves, a trend that hasn't improved since the initiation of the state's well-intentioned project.

One obvious drawback of the government's various interventions is that there is no system to record, with any level of professional skill or consistency, whether a farmer who died by suicide was suffering from depression, or any mental ailment, or even the kind of stressors that pushed him over the edge. In fact, the recording of farmer suicides is limited to whether the deceased farmer owned land and whether the land bore financial liabilities. This information is collected by the revenue department to determine whether a case is eligible for an ex-gratia payment. Common reasons cited for declaring a case ineligible include absence of loan or crop failure. Landless labourers who may kill themselves will not be counted as farmers. Women farmers who do not own land will not be counted.

The system rules out any possibility of nuance, so while crop failure and indebtedness may be among the most common reasons cited in farmers' suicide notes, varied mental health stressors cannot be studied. These could include market fluctuations, poor price realizations, inability to afford better healthcare for a chronic ailment for oneself or for a family member, or inability to afford better higher education for a daughter or son. As for somatic symptoms, these may simply escape all mention. Conversely, in cases where a farmer who died by suicide very obviously suffered from a mental illness with symptoms of psychosis or catatonia, farm-related stressors are sometimes ruled out as a cause.

Several cases of farmer suicides have been declared 'ineligible' for government compensation on the grounds that the cause was a pre-existing mental condition, not farm stress. In their lived reality, however, farmers' encounters with mental and physiological ailments, minor or major, and their socio-economic and financial conditions and interpersonal stressors are intricately intertwined, challenging simple cause-effect checkboxes that government officials expect to mark.

A more analytical study and a deep dive into the data generated by the Prerana teams has also not been done. Though there is no qualitative analysis of the data they generate through daily consultations and referrals, the Prerana team mostly agree on some similarities across districts. For one, 'stress is everywhere' in these districts, as one Prerana medical social worker put it, arising from the gamut of problems associated with unirrigated farms, poor rainfall, dry wells, indebtedness, failed crops, financial worries and the consequent domestic disputes. In addition, there are relationship problems and family discord, not always arising out of the financial stressors.

For the age group under thirty, doctors say, the internet is a common source of unrealistic objectives and goals as inexpensive data and social media bring to their mobile phones glimpses of wildly different lifestyles. Shinde of the Osmanabad Prerana team says dementia is on the rise, often appearing to be hastened by senior citizens' experience of living alone in villages emptied out by migration, though this inference is yet to be backed by data analysis.

The other thing most members of various Prerana teams agree on is that ASHA workers still have only a rudimentary understanding of the various terms, such as neurosis, psychosis, stress, anxiety and suicidal ideation. So, though they play a key role in spreading awareness about mental health issues in villages, they have been unable to identify potentially serious cases.

'If our teams of better-trained professionals were to reach every village, we might be able to show results, but there are limitations to how much a team of six or seven can do in large districts with about 1,000 far-flung villages each,' says Shinde. There is also the ubiquitous problem of vacancies – civil surgeons and psychiatrists at district hospitals are often not available because the post is still vacant. Also, the Prerana teams expect better results in preventing suicides once the low level of awareness regarding mental health is reversed and reporting of mental health problems becomes more routine. On that front, as stress and anxiety increasingly become tea-shop conversations, their work is beginning to show results as the stigma associated with mental illness shrinks gradually.

~

IN THE SUMMER of 2014, troubled by three farmer suicides in villages around Aurangabad, doctors and social workers at the Dr Hedgewar Hospital, a large multi-speciality facility in the city, which is the divisional headquarters of Marathwada's administration, put together a small intervention plan and also collected data to provide context to the unexpected suicides. The hospital had been conducting weekly or bi-weekly OPD clinics in a few dozen villages, mostly in the Aurangabad district. They were also operating mother-and-child clinics, a wildly popular outreach programme in these remote villages where speciality paediatric healthcare was unheard of until then.

The hospital is popular both as a medical institution and for its philosophies and outreach programmes not limited to medicine and preventive healthcare but also extending to water conservation and agronomy. So, when farmer suicides shook three villages where Hedgewar Hospital conducted regular outreach work, the doctors who helmed these OPDs began to sense that some advocacy on mental health was essential.

This was during the tail end of the 2014 drought. This was also an opportunity to study further their observations on addiction and substance abuse. Coincidentally, a doctor scheduled to join their team took his own life in his hometown, Pondicherry, around the same time. Dr Pratibha Phatak, project director of the hospital's rural healthcare unit, remembers being shell-shocked. The need to make mental healthcare available widely suddenly became that much more urgent. Weeks later, the rural healthcare team launched Arunodaya, or the sunrise project.

Through their project, they organized poster exhibitions in villages on stress, mental health, alcohol abuse and de-addiction. On diagnosing residents as requiring an institutionalized de-addiction programme or counselling, some were given therapy sessions by a trained psychologist while others were sent to Pune's Muktangan residential de-addiction centre. They also appointed local youngsters as 'Manas Mitra', or friend of the mind, to offer some counselling and to spread awareness on mental health in a handful of villages. Later, as the state government's Prerana Prakalp was launched, the Manas Mitra volunteers also underwent training provided by the programme coordinators and psychiatrists.

The efforts of the Arunodaya project and the Manas Mitras often focused on de-addiction, in consonance with the team's experience that alcoholism has kept pace with rising farmer suicides in the region, and that the linkages between the two need further study. 'Earlier there wasn't a single wine shop in the villages. Now there are four or five in every village,' says Dr Pratibha.

Consumption of alcohol has undeniably risen in drought-hit rural parts. Where consumption was once an event, for it required a ride to the nearest town or large village and a pre-planned outing with friends, a drink is now only a short walk away for most residents in Marathwada's eight districts. Recent years have witnessed bars and beer shops and dhabas where alcohol is served sprout even in medium-sized villages of the region. 'Bar *chalu aahe*,' is a common signboard across the highways, directing revellers to newly located bars. In most places, the establishments are not positioned along the highway, in deference to a Supreme Court directive. In some districts, where local government bodies successfully redesignated highways as local roads, the bars are close to the highway. Everywhere else, the signposts are on the highway, silent markers of a new way of living in the region.

Alcohol-related stress does not limit itself to the men consuming it. The Arunodaya project found through surveys in 2015 and 2016 that anxiety and stress were very high among women with a heavy-drinking man at home. Spousal conflict over drinking and consequent mental trauma were very common, sometimes including domestic violence and its concomitant anxiety.

The Arunodaya team's study dated May 2015 had found addiction to be prevalent in at least a quarter of the families surveyed. In 2016, they found this to be increasing slowly. A separate study on mental and behavioural problems among female farmers revealed that nearly half the participants were 'plagued with anxiety and significant amounts of depression and stress'. In detailed interviews with ninety women farmers, conducted in April and May 2016, in eight remote villages of Aurangabad district, 50 per cent were found to be facing anxiety, 70 per cent were under stress while 17 per cent were suffering from depression. In a cross-sectional descriptive study of 280 women farmers in eight villages of Aurangabad district, the participants ranging from eighteen to forty-five years, the researchers found over 77 per cent had a family member with chronic addiction. Sixty-five per cent reported extensive losses due to drought and 43 per cent reported indebtedness and related stress. About 20 per cent were worried about daughters of marriageable age. Significantly, over 8 per cent had harboured suicidal thoughts.

Stress due to the twin problems of drought and an alcoholic husband is related to landholdings too – larger percentages of those with less than five acres of land reported 'severe distress', while larger percentages of women with holdings over five acres reported milder mental distress, says Dr Madhuri Gavit, the team's psychologist. The study used structured questionnaires containing measures of psychological distress as per the Kessler Psychological Distress Scale. Interviews with participants were conducted in person and their responses written down.

They also found that women developed their own coping mechanisms, including casual abuse of over-the-counter medication and other substances. Women also suffered from distinctive, sometimes very subtle, kinds of trauma.

One winter morning in December 2017, Pandharinath Bhurye of Kamkheda village in Aurangabad, barely twenty-eight years old, went to the nine-acre farmland owned jointly by him and his brothers and consumed pesticide. He had taken a loan to dig a well that yielded no water. Vaishali, his wife, was only twenty-two then. Having resolved to send their son, now five, to an English-medium school in Phulambri town nearly fifteen kilometres away, the spunky Vaishali refused to hand over the farm for cultivation to another farmer. 'They were offering ₹35,000 to use it for the season. I decided to till the land myself. I invested ₹60,000 and made a ₹90,000 profit,' she says. But walking to the farm alone and doing the hard labour herself means the young woman attracts plenty of attention in Kamkheda. 'It is a kind of torture. The men pass lewd comments, or they hang around my house. I have learnt not to react at all; I just stay silent,' she says.

In nearby Anjandoh village, Sunita Soninath Shingare, also a farm widow in her twenties, says, '*Manus ha sagla kaahi aahe.* Your husband is everything.' Her husband's addiction had relapsed. He had stayed clean for a year with the help of the Arunodaya project's Manas Mitra volunteer before spiralling again when a goat farm he set up on credit failed. First one goat died, then another. Late in 2018, he went to his two-acre farmland and drank the locally available 'desi', a sweet, strong country liquor, lacing it with a common farm pesticide. He was dead when they found him. Sunita tills the land herself now — she cannot afford the expense of farm labourers anyway. But the land is not in her name, and she is always too afraid to borrow money to improve her prospects. 'People will want something in return,' she says. Her stressors are entirely different.

A one-size-fits-all approach can never be effective while counselling farmers fighting depressive bouts and loss of self-worth, or while reassuring their family members coping with the accompanying hazards. Vishnu Piwal, a young social activist in Jalna district, raised some money during the worst of the 2014 drought and trained eight social work students as counsellors, one for each of the eight talukas in his district. As a student of social work at a local college, Vishnu had been deeply affected by a friend losing his father to suicide in distant Yavatmal. So as soon as he was able to gather like-minded people and funds, he started the Phirte Shetkari Samopdeshan Kendra, or the itinerant farmer counselling centre.

'We were doing a lot of what the Prerana team eventually did. As early as 2014, we went from village to village, and tackled each area and each kind of distress differently,' he says. Having roped in doctors from Jalna's established de-addiction centre and mental health clinic, Manas Hospital, to train his young counsellors, Vishnu initiated a scheme that continues to function loosely even years after their counselling sessions ended. He set up a counselling advisory team in the villages his team visited – a teacher or principal, a senior member of the village community, a doctor if one was available. Crucially, he made sure each of the twenty-seven or twenty-eight such advisory panels he formed included an official from the nearest bank.

'That was key – because those getting notices from the banks about impending seizures of their property were the ones living in the darkest fear. To have a bank officer tell them they shouldn't panic made quite a bit of difference,' Vishnu says. The village advisory teams would identify the most vulnerable families and refer them to the counsellors. Often, farmers needed to know they had the support of their families despite the circumstances. Some needed financial literacy lessons. Some needed physical help with an ailing relative. There were almost as many different methods employed to mitigate vulnerability as there were distressed farmers the counsellors met. They eventually reached 1,100 farmers in eight talukas before winding up a little after the Prerana programme was launched. But some of those village advisory panels still function, the local seniors and professionals still sometimes spot a man in need of counselling and together convince him to get help.

Many of those who were counselled in 2015 do not want to talk about it, apparently embarrassed by their own frailty at the time. One farmer, having been helped by Vishnu to step away from the brink in Jalna's Mamdabad village, resolutely refused to concede later that he had ever contemplated taking his life. But on meeting him after a long journey to his little farm-side home, his wife was in tears as they recounted their saga of financial despair when 2014 and 2015 ended up being back-to-back loss-making years right after the 2012 drought.

The Arunodaya team also briefly ran a programme called the Kishor Panchayat in ten villages, working with 1,000 boys between ten and fourteen, sensitizing them on HIV/AIDS, adolescent health and their unique addictions including sex. High-risk sexual behaviour is one of the visible aspects of the changing social milieu in Marathwada, along with young boys consuming tobacco or cigarettes, doctors said. 'Young men in Marathwada don't seem to want to work in the fields any more,' says Dr Pratibha. The rewards are poor, when there are any. The hard labour under the harsh sun is a daily reminder of the futility of their ten to twelve years of school and college. She sees them idly sitting around, unemployed and unwilling to take on the poorly paying labour on other people's farms and the unpaid labour on their own family's land. They gather at the village square or street corners, playing games on their mobile phones, trawling social media or riding their bikes around aimlessly.

Dr Pratibha says the spate of government infrastructure projects, including roads and industrial parks, has brought its own peculiar problems, such as the driving of flashy new cars under the influence of alcohol. This is common especially in parts of Aurangabad district along the Delhi–Mumbai Industrial Corridor, particularly the Aurangabad–Ladsawangi road. The stories of men who go drinking in these areas have become something of an urban legend in Aurangabad – one tale is about a former farmer who recently sold his land to the state government for the DMIC project. He went drinking 'with ₹1 lakh cash in each trouser pocket' as one Ladsawangi resident described it. He refused to loan some cash to a friend, and a scuffle ensued. The man was found the next morning, fatally stabbed, the money gone. The problem is also that there is no financial literacy alongside the large payouts from government or private entities for their land.

'Ask the women, they'll tell you these years have led to men losing their *purushartha*,' says Dr Pratibha. Translated as the essence of manhood, the term also relates to morality and righteousness. 'Earlier men would say *jithe laathi maaren tithe paani aanen.* (Where I strike the ground, there I shall find water.) Now they just take their bikes and flee, whether it is for mobilization of party workers during elections or a morcha or something else.'

FIVE

ON THE ROAD

'अरे शेतात घालूनी कष्टाचा घाम, काळ्या मातीत पिकवतो सोना'
(Labouring in the fields, you shed beads of sweat,
reaping gold from black soil.)

—Lyrics from a song performed by tribal farm-labourer
Alka Dilip Mali

NOBODY IN CHARDARI village owns an almanac. But when they feel the first nip in the air a few weeks after the south-west monsoon has departed, right around Diwali, every household in the village located behind the hill slopes of Dharur taluka in Beed knows it is time to start packing. Two to three sets of well-worn clothes, basic kitchen supplies, a set of utensils, blankets – the sparse bundle of items tied in an old sari must suffice for four to six months, says Anita Dongre. The *mukadam*, or contractor, is scheduled to arrive any day now with a date of departure. Eighty per cent of Chardari's residents between the ages of eighteen and fifty will leave home, clambering aboard bullock carts, tractors and trucks, undertaking a journey of 300 to 800 kilometres.

Chardari's 400-odd families are among the nearly eight lakh people from across the state of Maharashtra who leave their homes in October–November for an annual relocation. Among India's largest instances of labour migration, these *oos-tod kamgar*, or sugarcane-cutting labourers, relocate for the cane-cutting season primarily to Western Maharashtra's lush sugarcane belt in Satara, Sangli and Kolhapur, where over a hundred factories process cane from 6,53,000 acres; and increasingly to Tamil Nadu, Karnataka, Telangana and Madhya Pradesh, sometimes spending three to six months of the cutting season well over 750 kilometres from home. Five lakh of these labourers belong to the Beed district alone, and of the remaining, the majority is from elsewhere in Marathwada. While some communities and castes have been traditionally occupied in the hard labour of manually cutting the stalks of the cane crop close to the ground, today, people of all communities, including those from higher caste groups, migrate every year for work as sugarcane harvest labourers.

Dongre had a couple of bad years financially, so when a *mukadam* was looking for workers to travel to Tamil Nadu in July 2019, she and her husband agreed immediately. They returned just a month before Diwali and were to again leave right after. One warm September afternoon, the neighbour's three-year-old son wandered in and sat on her lap. His right foot bore scars from their last trip when he stepped on the embers of a cooking fire, just one of the many hazards for this sea of labourers who migrate annually for work. Anita works right through her period, illness or injury.

A handful of government resolutions exist to assure these labourers of some security of minimum payment and insurance but, in effect, Marathwada's *oos-tod kamgar* are neither recognized as workers, a status requiring sugar factories or contractors to provide them benefits, nor even as contractual staff with some manner of formally documented work conditions and pay security. Unorganized and overwhelmingly unlettered, they still account for cash payments worth several hundred crores of rupees each year. It seems like a large sum of money but works out to an average of ₹70,000 per couple for four to six months of work, or ₹7,000 per person per month for twelve to fourteen hours of hard labour a day without leave, medical care, accommodation or food.

The work is physically onerous, each couple cutting three to five tons of cane a day. It entails hacking at the green leaves covering the cane stalk, then slicing the stalk, that may be two to four metres tall, close to the ground and trussing it up in bundles before finally loading it on to a truck or a bullock cart – all this for hours without a break. It is often dangerous: snakebites and insect bites are common, as are injuries from the slicing implements and machetes, trucks, cooking fires and wild boars and dogs in the dense foliage of the cane fields. For most of these workers, the reason to sign up is the opportunity to earn a lump sum of ₹50,000 to ₹1 lakh, otherwise an impossibility.

Through the winter and early spring, villages like Chardari are largely home only to the aged, disabled and the children left behind in their care. And as cane fields in Tamil Nadu and Madhya Pradesh appear to offer marginally better prospects, these hardy men and women, very often accompanied by their children, migrate farther and farther away to make a living.

Now in his sixties, Sampatti Shamrao Munde of Chardari remembers when cane fetched ₹2.50 per ton harvested. And because a husband–wife couple would be able to cut and transport just about 800 kg to one ton, that was the day's pay, more than forty years ago. 'I've never missed a year, for at least forty years. First, we used to go to Dongarkada in Nanded, then we began to go to Latur, and now for three years I've been going to Karnataka,' he says. His left leg has suffered some atrophy, and normal movement in it is now impaired. But he and his wife, almost the same age, will keep making the annual journey. 'We'll stop when one of us is completely disabled or so sick that we cannot move,' he says, his eyes expressionless under the little clouds of cataract.

The journey, to Western Maharashtra's Sangli or Satara, or Nanded further east in Marathwada, or neighbouring Latur, used to take a couple of days to a week. The families would set out on bullock carts, wooden carts with large wheels yoked to a pair of animals that ambled on all day, stopping only for one sparse meal during the day and to spend the night by the side of the road. It was only in the mid-1980s that the 'tyre-*gaadi*' arrived as an important innovation in the industry – a metal trolley or cart that could be yoked to bullocks.

According to senior *mukadams* who have been in the industry for decades, it was when Jayant Patil of the NCP became an MLA that the tyre-*gaadi* began to grow popular. This was safer: the prefabricated trolleys did not break down so easily on rutted roads, and they made for more comfortable journeys. But there was one more reason for the innovation to become rapidly popular. Where a wooden cart yoked to bullocks allowed a couple to transport at most one ton of cane a day to the factory, located anywhere between four and forty kilometres from the field, a metal trolley could hold two to three tons of cut and bundled cane stalks. It was progress for the workers and for the factory owners – more work, larger earnings, more cane every day at the factory, optimum usage of machinery, more capital investments and overall increased productivity.

Once in the cane fields, the labourers erect their camps where they will live through the cane-cutting season. The home is called *kopi*, a simple tent made of bamboo poles and a large cloth, plastic sheet or tarpaulin, a discarded sari or other fabric sometimes added as a partial layer on the interior. Some *mukadams* provide stretches of thatched material, sometimes at a cost that is deducted from the labourers' pay. The *kopis* are erected in a clearing in the field, never inside a village. Inside the *kopis* are stacked the four or five kg of millet or sorghum meal in cotton bags or tin boxes, rolls of bedding and threadbare blankets to ward off the biting ten-degree cold at night and the mosquitoes. Scythes and knives are hung on a peg, but still within the reach of children. The floor is usually a piece of tarpaulin on the prickly ground.

There is often no source of drinking water in the vicinity, the nearest well may be a couple of kilometres away. Sanitation, even in the form of a basic hole-in-the-ground toilet, is almost always entirely missing. Food is cooked on a rudimentary stove made of stones, fuelled by firewood and anything else that can be collected and will burn. Who visits the *kopis*, who accepts a cup of tea boiled on the stones of the outdoor hearth, whether a woman in the *kopi* gets a male visitor – all of this is controlled by intricate power dynamics in the camp. Many have heard of women being sexually exploited, but in the absence of police complaints and verified statements, it remains in the realm of camp gossip.

The women know things are doubly tough for them, having to run an entire household from a *kopi*, turning the cloth and bamboo tents into something resembling home for the family, making sure the children are fed and watched over when they are asleep, making palatable meals with whatever supplies and cooking implements are available. 'So many times, we have had to collect water from the gutter alongside the field,' says Shyamal Dongre, a fifty-something woman in Chardari.

There are no set working hours. If the factory is crushing late into the night or in the early morning hours, that's when bullock carts, trolley carts and now increasingly trolleys and twin-trolleys pulled by tractors begin to queue up at the gates. That means, Shyamal says, sometimes they leave their *kopi* at 4 a.m., or they return from the factory at 2 a.m., routinely clocking sixteen- to twenty-hour days. 'Depends when the vehicle comes,' says Ankush Munde of Chardari. 'Sometimes we call the driver to tell him the *maal* (cut and bundled cane) is ready, but there's a queue of vehicles at the factory and the driver comes several hours later. Nobody has ever come up with a system to streamline this.' If they are leaving for the factory at night, their children have to be left behind, under somebody's watch. If they are cutting cane at some distance from the factory, anywhere between five and fifty kilometres, then the travel time is added to the wait time. 'Some of these things have become more difficult even though there's better technology and vehicles available,' says Ankush.

Other things have remained the same for decades, such as the system and mode of payment, the record-keeping, the hazards the labourers live with.

A payment in advance, called *ucchal*, literally translated as 'lift' and ranging between ₹50,000 and ₹1 lakh for the season, is made, always to a couple. A man–woman pair is always a single unit of labour, called *koyta*. The *koyta* is actually the delicately curved sickle used to hack the cane stalks, and a husband–wife pair is called that and paid as a single unit because they often work in tandem: one cuts while the other strips the excess foliage and ties the bundles of felled stalks. Children help too, even those as young as seven or eight years, and they are entirely unpaid.

Surekha Chavan, now seventeen, of Ramwadi village in Jalna accompanied her parents for several years until a local NGO began to assist in overseeing her education while she stayed back in the village. Quick-witted and full of laughter, she is amused when asked to describe her life in the *kopi* and in the cane fields. 'We would help cut off the leaves – the kids could take a sickle and hack those off,' she says. She never got paid, but the occasional ₹10 note folded into her palm sometimes by a *mukadam* or her father was always a happy treat. Surekha is now trying to complete her education, helped by an organization called Krantisinh Sanstha, but the pressures are manifold – girls with degrees do not have better marriage prospects in the community unless the parents can afford a large dowry; there are no jobs locally for Surekha if she completes Class XII and graduates. Marriage among teenagers is common because the newly married couple becomes another productive unit on the cane fields.

Despite interventions from NGOs and despite legislation, child marriage remains common among communities that work on the sugarcane fields. Shesherao Munde of Chardari laughs heartily when he is reminded of how he hurriedly got his teenage son married a few years ago, the girl no more than fourteen then. The family was going through a difficult patch and the prospect of another *koyta* was tempting, for it meant one more *ucchal*, one more lump-sum payment in the household, he says sheepishly.

For those with better hopes for their daughters, it is not a matter of amusement. For the last few years, Ankush Munde was making use of a government scheme to keep children of sugarcane labourers in school during their migration, by ensuring two meals for them every day in the premises of village-level schools run by the zilla parishad, the district councils. In 2006, following sustained advocacy on the problem of cane workers' children skipping school for months, the Maharashtra government launched a hostel scheme for parents who wanted to keep their children back at home while they were away in the cane fields. Though a government resolution on the seasonal hostels, or *hangami shalas*, was passed in 2006, actual implementation began after the Right to Education Act was passed in 2009, making it mandatory for all children to be in school. Under the government resolution, seasonal hostels were to be started for all villages where twenty or more children migrate with their parents. Keeping them back instead, the hostels were to provide food, health services and education. While the scheme and the seasonal hostels continue to exist, there are only a few dozen operational.

The father of two boys and two girls, Ankush managed to get one daughter married a couple of years ago – to a family that does not engage in cane cutting, one of his life's biggest successes. His other daughter is now in Class XI, and he's keen to see her complete at least Class XII, but was unable to keep her in college during the academic year of 2019–20. 'She's grown up now, we cannot really leave her behind with anyone. Who can we trust?' At seventeen, the girl was a seasonal migrant along with her parents that year, returning in time to take her exams. Everybody recollects instances of young girls being sexually assaulted or exploited in the absence of their parents, and Ankush's decision won murmurs of resigned approval.

'*Hangami shalas* are only useful for contractors who run them,' says Datta Munde, a contractor in Chardari. His four children, including two daughters, are in Beed city, where the family now lives, while he makes the thirty-kilometre journey to Chardari a few times each week. 'We're educating our children, but where are the opportunities for them anyway? Even after doing her Class XII, Ankush's daughter may well end up getting married into a family of sugarcane cutters. Our sons face the prospect of living exactly like we have for all these years. We would like our children's generation to do something better.'

Vishnu Pival of Krantisinh Sanstha in Jalna says if the seasonal hostels worked effectively, girls like Surekha would have had a much better education. Pival's team of women and their self-help groups now also look after girls like her who stay back in the village during the cane-cutting season, and give them vocational training in embroidery or sewing or a basic course on operating a computer.

In Khodas village of Dharur, Suman Baburao Pawar chose to stop working as a cane labourer a couple of years ago, after doing it for the better part of a decade. Suman's daughter, Bhagyashree, is in Class XII, studying in a college in the neighbouring village of Adas. Bhagyashree has to walk two and a half kilometres to the main road from where she can get a bus to Adas. The family used to leave her behind, taking the elder sons along. 'But she's now at an age where we cannot leave her with anyone,' says Suman. Even in villages where crime rates are low and violent crimes are rare, young girls' safety is difficult to guarantee. Suman works as a farm labourer in Khodas now, but work, which pays anywhere between ₹100 and ₹150 per day for a woman, is not available all round the year. 'Now we're harvesting soya bean, but if there's no rain one year then there's no crop, and so no work here.'

About her own future plans, the seventeen-year-old Bhagyashree says she has none. Studying further is not practically an option because of the difficult commute to Adas and a lack of company. She hangs out with her cousin's wife, Lalita, and gets a dim idea of what her next few years may be like. Married in 2018, Lalita appears barely eighteen years old. She used to accompany her parents to Western Maharashtra for the cane-cutting season and will soon accompany her parents-in-law, husband Avinash, his two brothers and their wives to Karnataka.

In the absence of an *ucchal*, Bhagyashree's wedding is going to require large borrowings. *Hunda*, or the bride's dowry, leaves most fathers poorer by a few acres of land if they have any, while others take large loans in the hope that a sizeable *hunda* will augur a better life for their daughter. 'If we want a doctor for a son-in-law, *hunda* is about ₹51 lakh plus hundred *tolas* of gold,' says Shesherao. A *tola* equals ten grams. 'For a teacher, it is ₹10 lakh plus some gold. Those who have land sell it off so that their little girl does not also end up cutting cane all her life. Those who don't have land, well, prospective grooms' families would just ask us to look for a match in a poor family like our own,' he adds bitterly.

The *ucchal* is paid by the contractor to the labourer *koytas*, all in cash. Not all contractors take advance payments from sugar factories, however. 'There are better profits if we put in our own money first and then collect payments later from the factory,' says Datta. His workers are paid ₹239 per ton of cane cut, loaded and transported, and an additional 25 per cent of that sum is paid to him. But without an advance, he gets paid ₹400 per ton at the end of the season. He maintains no legally enforceable contracts with his workers, nor with the factory. All records are maintained in a pocket notebook. The labourers now all have Jan Dhan bank accounts, but the cash advance is rarely deposited into one. Mostly, there are waiting expenses such as children's college fees, a wedding *hunda* or a well to be dug in their small landholding. Contractors work entirely on cash, they say, because that is how the factories operate. There is no tax invoice or records for payment of state taxes. And because contractors work on a 10 to 25 per cent commission, paying 12 to 18 per cent in taxes would not be affordable.

'The system works on trust,' Datta says. For example, an advance of ₹1 lakh is paid to a *koyta*, but in a bad year – and there are many – a couple manages to provide labour worth only ₹60,000. 'The next year's advance is then lower.' The system, mostly unchanged for years, also serves to keep labourers perpetually in debt to their contractors, for there simply may not be enough cane crop, and a *koyta* is often unable to work out the entire advance. Labourers are full of stories of other ways in which this system is exploitative. There is no proof of how much they were paid or penalized or how much more labour they owe.

But Datta says the absence of any formal record-keeping is equally dangerous for the contractor class. He remembers a case a few years ago when a woman labourer fell from a raised platform and was seriously injured. He says he paid ₹90,000 in hospital bills, but she had lost so much blood in the first couple of hours that she could not be saved. 'I had to bid goodbye to the rest of the advance I had paid. The husband never returned to work, and the money I spent on her hospitalization also had to be written off.'

A social security scheme for sugarcane labourers has been on the cards for decades, including proposals for government-backed loans for the workers. But after years of promises by consecutive governments, the Loknete Gopinath Munde Maharashtra Sugarcane Harvesting Welfare Board was finally set up just days before the model code of conduct kicked in ahead of the 2019 state assembly election, announced by a government resolution on 13 September. A former legislator from Beed, himself a former cane labourer, was announced as the head of the welfare board. Oddly, though, the announcement of its formation came with the rider that all other details, including the functioning of the corporation, its finances and its schemes, would be decided in due course of time. Meanwhile, almost all major sugarcane *mukadams* believe the government should legislate to protect them. With a regime change by November 2019, this plan was relegated to the back burner.

~

HANUMANT NAGARGOJE, FIFTY-FIVE, is one of Beed district's bigger contractors, with about 5,000 labourers. But 'Bappa' Nagargoje, as he is referred to in Dharur, also an elected representative to the Panchayat Samiti and former elected member of the Beed Zilla Parishad who lives in a palatial three-storey bungalow in Dharur town and drives a luxury SUV, was once a sugarcane labourer himself. 'I have seen both sides of this life,' he says. 'For fifteen years I was a labourer, cutting cane with my wife in Western Maharashtra and then in Karnataka. Growing up, my family did not have money to buy foodgrain, so we would forage for wild vegetables. Today, I live like a king.'

Hanumant's is a rare rags-to-riches story in the ranks of the labourers, having risen as a contractor to become a taluka-level politician of note over twenty-five years, now earning commission from factories for the 5,000 labourers, 200 trucks and 300 bullock carts that he provides.

Over forty-five years ago, a young Hanumant started off working as an agricultural labourer. His family owned a small piece of land, but it was unirrigated and yielded no crop. The family worked on other people's land. The youngest of four brothers, he would tend to people's sheep and goats for a living. He made the decision to join the cane-cutting migration as soon as he got married, keen to establish an assured source of income. He spent his first cutting season in the fields near NCP leader Jayant Patil's sugar factory in Islampur, in Western Maharashtra, about 350 kilometres from his home in Dharur.

He moved from one factory to another, one labour contractor to another, working in Tasgaon and Kavthe Mahankal in Western Maharashtra before travelling to Karnataka. It was when he was cutting cane in Tasgaon that he met a relative who was friendly with the vice-chairman of the factory there. His relative put in a good word and Hanumant was suddenly on his way to becoming a *mukadam* supplying labourers on five bullock carts. He had to take the family's bullocks and hire a few more to meet the requirements, but continued to cut cane too, alongside his wife. Very slowly, over nearly a decade and through the help of relatives and friends, his work ethic was noticed, and he began to get opportunities to provide labour to factories such as Jamkhandi Sugar Factory in Bagalkot and later in Davangere, both in Karnataka. Hanumant now earns a ₹10,000 commission on every ₹1 lakh paid by the factory to labourers, and 2,500 *koytas* working under him get an average seasonal pay of ₹75,000.

One of his sons is now with a central paramilitary force, while the younger one will take charge of a business set up especially for him, most likely a medical drugs store. Hanumant has a special introduction for his son, a strapping lad who's recently completed a degree in pharmaceutical sciences. 'He was born in a *kopi* in a cane field. Right after my wife finished loading two trucks with the cane that we had cut that day.' Heavily pregnant, she continued working even after her water had broken, he says.

Hanumant Nagargoje is an imposing presence in a room, wearing a kurta and loose pants, both spotlessly white, and a white Gandhi topi that does not budge from its position even when he gesticulates animatedly while speaking in his booming voice. He stands almost six feet tall, and there is an imperturbable air about him. But his eyes moisten as he remembers that day in Sapsagar village in Karnataka's Atni taluka where they were working. 'Everybody ensures women who have delivered a baby get at least forty days' rest. On the twelfth day after she delivered, I took my wife back to the field with me where she helped tie up the bundles. Our eleven-day-old son was right there by our side, lying on a cloth on the field.' He says his success stems from the humility and integrity that the hardscrabble life stamped on him. 'I know the experience and I remember it vividly, so I never take what is rightfully a labourer's due. I earn from the factory's commission, and not an extra paisa over that by cheating the workers.'

A few years ago, when he was working as a *mukadam* in Latur's Killari, a bank official in Umarga who had welcomed him warmly upon seeing his appearance and clothing eventually refused to give him a bond he needed because he couldn't sign his name. 'That day I took an oath that come what may, I will ensure that my children are properly educated.'

Formerly a member of the BJP, he joined the NCP around 2013–14 and has since been elected to the Panchayat Samiti and the Zilla Parishad. 'I sleep like a king. I have a ₹5 crore estate. I constructed a three-storey bungalow for my family in Dharur, a two-storey home in the village, a four-storey building nearby, another two buildings are under construction. I drive a ₹25 lakh car, my manager uses a ₹9 lakh car. One son drives a small car, the younger one owns a Bullet motorcycle worth ₹1.5 lakh. From *shunya* (zero), I built my own mountain,' he says.

A powerful man who walks around with large wads of cash tucked into the waistband of his trousers, Hanumant concedes that the absence of formal contracts with labourers renders them helpless. As a former labourer himself, he is quick to acknowledge that a series of reforms in the system could improve the lot of labourers. And though he is associated with the NCP, he says no party in government has ever applied its mind to resolving the cane labourers' problems. At the same time, factories have begun to pay ₹450 to ₹500 per ton for cane that is cut by machines while labourers who risk life and limb and get bitten by snakes and poisonous insects while working in the fields get paid only half that much.

According to him, the newly formed welfare board for cane workers should make it mandatory for factories to pay insurance premiums for the labourers or make the payments itself. Workers should be given the status of employees with an annual or monthly salary, he says.

But for all his sympathy for the labourers, Hanumant says a longstanding demand from the association of contractors in the sugarcane industry – a law to protect them – must be urgently accepted. 'Sugar factories take all kinds of collateral from contractors if they pay us a large advance, from our farmland to our homes and gold. They have all our details in their records, including our PAN and our Aadhaar details. But while our assets are kept as security with the factories, the same is not true of our relationship with labourers – the advances are given and honoured entirely on a personal relationship with nothing in writing. I can tell you dozens of cases in which *mukadams* have had false cases slapped against them by labourers who have not worked off their advance payments. We take no security from them. All we do is buy a little diary worth ₹5, note down in it the names of the workers and a date of payment. It can never be used as proof of anything.'

He says *mukadams* have faced cases under a law preventing usurious moneylending and also under a law meant to protect members of the scheduled castes from violence and discrimination by upper castes. 'In our village there are members of various lower castes. I address them as *dada, tatya, kaka, aaba,* always with great respect. But that won't stop them from filing a case against me under the Atrocities Act [Scheduled Castes and Tribes (Prevention of Atrocities) Act]. We need to have a rule that some proof is required before cases are registered against us about using casteist slurs while talking to them,' he says.

~

VANJARWADI, OR A hamlet of the Vanjari community, is sometimes referred to locally as the village without wombs. At least forty women in the little village of about 250 families, between nineteen and sixty years of age, have undergone a hysterectomy. In 2018–19, as it emerged that a large number of women from Beed district engaged in sugarcane cutting had complete hysterectomies, a flurry of half-truths, loose allegations and wild misinterpretations followed.

As Vanjarwadi's women confirm, instances of surgical excision of their '*pishvi*', literally translated as bag, have been common for over a couple of decades. The decision for surgery is often made amid a haze of poor medical care, poor sanitation, a profiteering doctor-surgeon-clinic nexus and fearmongering about cancer. But some reports made the somewhat simplistic claim that the hysterectomies were purely on account of *mukadams* being unwilling to hire menstruating women, and the matter quickly snowballed. Women's groups called for an enquiry, the issue was discussed in the state's legislature, and an expert committee was set up under no less than the deputy chairperson of the Maharashtra Legislative Council to investigate the circumstances around the unusually large number of hysterectomies in Beed. The panel is expected to formulate guidelines for private clinics conducting such surgeries. In the committee are three women legislators and three medical experts.

Pankaja Munde, former legislator from Parli and minister for women and child development in the BJP-led state government between 2014 and 2019, rightly pointed out that women undertake hard labour in various sectors including other kinds of farm work outside the cane fields, and Beed's hysterectomies would need to be studied in greater detail. 'The department will make an enquiry into all aspects of this matter – the age group of these women, their specific problems, etc. This is a migrating population and they are all labourers. This is a new fact that has emerged about them, and if a migrating working population is taking such a step to strengthen themselves for work then it is a very sensitive issue,' Munde told reporters in June 2019.

But a preliminary investigation by the civil surgeon in Beed pegged the number of hysterectomies across ninety-nine hospitals in the district at 4,161 between 2016 and 2019, a minor fraction of the approximately four lakh women engaged in cane-cutting work across the state, and two and a half lakh in Beed district. And in Vanjarwadi and elsewhere, women who had the surgery suggest that a complex web of factors led them to the operation theatre, the pressure to work without rest in the cane fields only one of them.

'In fact, I ended up slowing down after my surgery. And a few years later, I began to have this terrible pain all along my left side, and now I can barely work at all,' says Lilavati Prabhakar Sanap, almost fifty years old. She had a hysterectomy twenty-seven years ago, at a government hospital in the neighbouring village of Raimoha. At the time, she was told there was no other way to cure her excessive menstrual bleeding but to surgically remove the uterus. Lilavati still goes to the cane fields each year, and looks for work as a daily wager the rest of the year.

The women's stories in Vanjarwadi have a common trajectory – of pain, excessive bleeding or infections during their period, likely from the absence of any sanitation amenities both at home and in the fields. Lilavati Rajendra Sanap was thirty-five when she began to get her period every fifteen days. She had a hysterectomy twenty-two years ago. Her daughter, married in a nearby village, was twenty-eight when she also 'got the operation done' about a decade ago. A young woman in her twenties requiring a medical termination of pregnancy ended up having a hysterectomy after doctors convinced her that it was necessary. Ranjana Govind Palwe, in her mid-forties, had her uterus removed about ten to twelve years ago after a bout of severe abdominal pain. 'I had medicines twice and one sonography, and they told me it was best to get the operation,' she says.

None of these women faced any direct pressure from a *mukadam* to get the surgery, but each one of them was made to believe she may get cancer if she didn't. Doctors, neighbours and other village women told them about women who had died of cancerous tumours in the uterus. None of Vanjarwadi's women was able to access a second medical opinion before opting for the surgery. 'Where's the time for that? We have to go back to the factory six months later, so it always seems best to quickly do what the doctor suggests while we're here at home,' says Ranjana. None of Vanjarwadi's forty women who had hysterectomies has any medical records or hospital discharge papers. And most underwent surgeries at the same two hospitals in Beed city.

The surgeries are also a financial blow for the women, sometimes setting them back years. Shafia Akbar Pathan, in her forties, had a hysterectomy eight years ago after suffering from excessive bleeding for a few months. The bill was ₹15,000 and an additional ₹3,500 for medicines. Unable to raise that sum herself, Shafia took an early *ucchal* from the local *mukadam*. Then, four years ago, she borrowed ₹4 lakh from another *mukadam*, for her daughter's wedding. The interest rate is ₹3 per month for every

₹100. That is an annual interest rate of 36 per cent, though she doesn't understand it, or doesn't know that there could be less expensive credit available. 'I'm still paying off that sum. The interest amount keeps growing, I have been working for smaller and smaller sums each year because I haven't repaid my earlier outstanding,' she rues. At her daughter's wedding, she gave the groom one *tola* of gold and ₹25,000 in cash. The girl and her husband are now both cutting cane as well, and live elsewhere in Beed district.

Kaveri Nagargoje, who, along with her husband, Deepak, runs Shantiwan, a home for children from families facing various kinds of distress on account of the agrarian crisis in the region, says the link between Vanjarwadi's hysterectomies and the sugarcane industry is deep and latent. Located in Arvi village, not far from Vanjarwadi and Raimoha, Shantiwan works with the women and men of the region, including the large communities of sugarcane cutters. Among its 1,000 school students and 300 hostel residents, many are children who would otherwise have been helping a parent cut cane.

'Even though there are now Swachh Bharat toilets in almost every home in Vanjarwadi, how do people who do not have water to drink use and keep a toilet clean? Across drought-hit villages here, the toilets lie unused while people continue to defecate in the open. And during the cane-cutting season, women have absolutely no access to any kind of sanitation; this is a daily source of humiliation and often leads to sexual assault. Getting their period in such circumstances is even more indignity piled on them,' Kaveri says.

Vanjarwadi's women do not use sanitary napkins, neither do the women in Chardari, Khodas or Adas in Dharur. The women all keep a length of an old sari or other cloth, cut into strips and washed and reused dozens of times. Kaveri says infections on account of the poor hygiene are common, and those are not treated properly or expeditiously. The problem is multifaceted: compelled by poverty, drought, lack of irrigation for their own farmland and absence of any other kind of salaried work, women are forced to migrate and live in unhygienic and hazardous conditions. There they are vulnerable to disease and exploitation and get no medical care upon contracting an infection. And when they fall sick at home, there is a searing pressure to get fit in time for the next cutting season.

In Chardari, women labourers are outraged when asked if there is ever a doctor available. In her fifties now, Shyamal Sanjay Dongre gave birth to a son inside a *kopi*, after a difficult last trimester during which she worked with swollen feet and an aching lower back almost all the time. 'It was a relief when I went into labour. I could stop working for a while,' she laughs shyly. She did not take a single day off during her pregnancy – a day off would have meant money docked from her pay. 'That day, I finished what I was doing in the field and then when the labour pain started, I went into our *kopi*. Three women put their sickles down and came with me. I delivered Vilas within the next half hour. Somebody cut the cord with a blade,' she says.

Vilas, grinning at being told of his own fuss-free arrival into the world, says he spent a few years as a sugarcane labourer too, but has now graduated to driving a sugarcane truck. He took lessons from other truck drivers and once he passed the driving test and procured a motor vehicle licence, he was able to make some changes in his life. For one, there's a fixed pay of ₹10,000 per month from the *mukadam* during the season, paid in advance. And when *mukadams* are working at more than one site, there is usually steady work for five to six months. But most importantly, his wife and children are safe at home in Chardari while he is away in Karnataka.

Vilas works for Datta Munde, one of Chardari's *mukadams*, who owns four trucks. His job is to be in touch with the labourers and arrive at their location once the sugarcane is cut and bundled, and is ready to be loaded. He then takes the vehicle to the factory. Accompanied by a couple of labourers, he inevitably waits for his turn with the tractor's music system blaring. There are only 1990s Bollywood songs available, but played loudly enough, it can be cheery, he says. Overall, it's much safer work than living in the fields and hacking and slicing cane stalks all day. But the roads are bad sometimes, and a six-tyre truck with trolleys attached to it can be unwieldy and could topple over. And there have been instances of a sleepy labourer falling off a truck and being run over. 'Still, this is better,' Vilas smiles.

Women have no such means of career progression in the industry – there are no women drivers or women *mukadams* or even staffers in the factories. Even if half the labour force here comprises women, the industry remains under the control of men alone. Working through their periods, knowing there is no emergency medical help, running a home and tackling the multiple problems of leaving behind an old mother-in-law or relative at home, it is the women labourers who get the short end of the stick.

In Chardari, Anita and Inder Dongre and their son, Tushar, now in Class XII, prepared to head to Karnataka for work after Diwali in 2019, having returned only weeks back for a short festival break with their families. Anita had to prepare a stockpile of essentials, pack up the simple cemented home carefully and check on an ageing relative. Sagarbai Dongre, almost seventy, lives alone during the months that the extended family is away in the cane fields. Her husband died young and her daughters are married in other villages. Her only son died a few years ago. The extended family's support is central to her daily life. She needs help bringing home water from the community tap, and she needs somebody to accompany her on hospital visits. Left behind during the cutting season, Sagarbai feels like she ages rapidly living alone, as days pass with little or no conversation to make. 'What's to be done? Can you just stop living? I make do. It's embarrassing when I have to ask villagers for money, but tell me, can you stop living?' she says, squinting to see clearly.

Sagarbai worked in the cane fields since her teenage years. '*Akshad taakla ki karkhaanyala,*' says Sagarbai. Akshad refers to the grains of rice showered on a newly married couple as blessings at the end of the wedding rituals. 'Once the blessings are received, it's time to head to the cane fields.' Many decades later, it is still the reality for her daughters and granddaughters.

As 2019's festival season approached, the workers prepared to leave; some had begun their journey already, from Dharur and Wadavani talukas to Karnataka and from Ashti and Patoda talukas to Kolhapur, Sangli and Satara in Western Maharashtra. The 2019–20 cane-cutting season in Western Maharashtra was to be short in some factories, only a little over two months, for thousands of acres of cane crop were destroyed by the floods in the region in August 2019.

Hanumant Nagargoje rues that a tradition that began in the 1960s when the Vasant Dada Sugar Factory was launched in Sangli has evolved with time but not adequately. Workers were paid ₹400 or ₹500 as *ucchal* in those years, or ₹7 a ton, but working conditions have barely changed. Instead of a bullock cart that carried 800 to 1,000 kg of cane, workers now use the '*chakri gaadi*' or the tractor-run six-wheel vehicle. One *koyta* accounts for up to eight tons a day. 'It's a manner of progress,'

Hanumant says, 'but nature has not been supportive. Those who purchased these vehicles have run into losses. They purchased a vehicle for ₹4 lakh, spent another ₹4 lakh on hiring labourers. But there isn't that much cane in a bad season, and earnings may be only ₹2 lakh. We see instances of finance companies seizing these vehicles, which makes it even more difficult to recover work from those paid in advance.'

Meanwhile, in Dharur, a few dozen storage tanks or artificial lakes built by the late BJP leader Gopinath Munde dot the hillside. These brought water for irrigation to some villages and introduced the concept of horticulture and vegetables. People began to grow tomato and onion; some even grew cane. Chardari had 2,000 tons of cane in 2018, down to 150 tons in 2019 on account of dipping groundwater and absence of rains. 'But overall, people have begun to eat two square meals, and the incomes from their farmland are able to somehow support their earnings as cane labourers,' Hanumant says.

But not even the marginal irrigation facilities provided by the string of storage tanks has offered an alternative. The region largely sees dryland farming, and while there is a large potential for the milk industry, it remains subpar on account of pricing problems and poor support to dairying. No other work is available, of course. 'We need the cane region to be rich and prosperous so that factories function for six to seven months every year. Otherwise, losses will pile up. Every labourer today finds he has been paid advances in excess of work done. A good cane crop every year is critical to their survival.'

In these villages, the euphemism for the migration to the sugar factories or *karkhanas* is to say the family is away 'in Belapur'. In Chardari, 80 per cent of those between ages of eighteen and fifty years 'go to Belapur' or migrate for the cutting season. Thirty per cent of those between fifty and sixty-five years also migrate. In addition, about 5 per cent travel to Mumbai, Pune or Nashik, especially those belonging to the scheduled castes who often own no land.

As communities sink deeper into impoverishment, labour becomes cheaper. Datta Munde of Chardari has begun to hire workers from the Ghansawangi area of Jalna district. They are willing to work for an advance of only ₹50,000 for the season as opposed to the ₹80,000–1 lakh paid to labourers in Beed. Workers are also prepared to go farther and farther away from home. Tamil Nadu and Madhya Pradesh are common destinations in the cane season now, and Karnataka with its irrigated cane acreage is a staple for Beed's workers.

There are some hopes from the newly formed welfare corporation for sugarcane workers. Truckloads of workers travelled to Mumbai to place before government officials their demands from the body – loans for tractors and daughters' weddings, mandatory accident insurance and more. Whether it will be a bipartisan body and whether the higher echelons of government ensure its effective functioning remain stories for a future date.

Everybody agrees that a better work life would be welcome, one that does not entail being away from home for long months. And everybody concurs that Marathwada's cane labourers need work in their home districts. As Datta Munde says, 'But nobody can solve this basic problem of bringing work to an area that has no irrigation and no water for industries. And so Marathwada will continue to have this curse, of eight lakh people migrating every year to cut sugarcane hundreds of kilometres away from home. But it is an essential curse.'

A HISTORY OF EXCLUSIONS

~

'Thus they create
A society based on inequality,
This being the inhuman ploy,
Of these cunning beings.'

—From 'So Says Manu', by Savitribai Phule (1831–97)

IN LAKHANGAON VILLAGE of Osmanabad district's Washi taluka, by the early 1990s, thirty families pulled out of the annual migration to sugarcane fields in Western Maharashtra's Satara and Kolhapur districts, where they used to work as harvest labourers for four to six months of the year. All landless Dalit families, they responded to a call made in the late 1950s and early 1960s by firebrand Dalit leader Dadasaheb Gaikwad for land redistribution. Through the 1970s and 1980s, these families began to occupy and cultivate community-owned land on the outskirts of Lakhangaon. Each of the thirty families occupied exactly two acres, marking off their boundaries with rows of stones. At the time, the land was overgrown with weeds and brushwood. They cleared the land, flattened it and ploughed it, the men and women doing all the labour themselves. As they harvested their first-ever farm incomes, their dependence on a lump-sum annual payment from cane labour contractors began to shrink. Those who remember the days of the occupation and the first harvest say it was a festive atmosphere when the first sacks of jowar from their own farms came home.

Much as it was their first brush with a stable livelihood, seizing government-owned land was not merely a source of income for the landless Dalit families of Lakhangaon. It was a satyagraha, an act of defiance. It was a show of solidarity with the tens of thousands who had participated in caste struggles in Marathwada since Independence. Bhaurao Krishna 'Dadasaheb' Gaikwad, a leader from Nashik, had grown in prominence as a close associate of Dr Babasaheb Ambedkar in the 1930s, during historical protests to demand that Dalits be allowed to enter temples and drink from public wells. Led by Gaikwad, the picketing of the Kalaram temple in Nashik lasted months and drew support from Dalits across regions. Then, in the late 1950s, a statewide movement was launched seeking land rights for Dalits, Adivasis and other landless communities, led by Gaikwad of the Republican Party of India and several communist leaders. Thousands from Marathwada joined the jail-*bharo* satyagrahas in 1959 and 1960, during which they courted arrest at various places. Around the same time, across Marathwada, Dalit rights activists had begun to refuse to do '*maangki*' or '*maharki*', menial work that was assigned as traditional caste roles. A new consciousness was emerging – of how land ownership, a continuing tool of caste-based oppression, was central to the rural economy.

In villages such as Lakhangaon, accounts of those struggles passed down through generations. So, Tai Survase understands that Dalit families' move to occupy and cultivate the '*gairan*' lands, or grazing pastures, on the outskirts of villages was closely tied to their self-image. These were all unirrigated plots of land, fallow and often hilly or rocky, hardly desirable for agriculture. And yet, between 1978 and 1990, according to the Maharashtra government's records, more than 1,08,000 hectares of land were occupied by landless Dalits across the state, a large portion of them in Marathwada's districts. Tai Survase was still in school when the last of the Lakhangaon occupation was taking place. 'We hired an earthmover to remove the trees on the plot. The families who were to till that land worked all night, the men chopping up the felled trees and the women stacking bundles of wood as others stood guard,' Tai remembers. It was all quite heady.

Led by a handful of organizations, Dalit families continued to occupy *gairan* lands through the 1990s. From 2000 onwards, a year before Gaikwad's birth centenary, thousands of families tilling *gairan* land in Marathwada began peaceful protests and demonstrations to gain lawful title to the land in their possession. They have made little progress two decades later, despite the state government on at least two separate occasions promising to convey the land titles to the occupiers' names. 'There is, in fact, no political will from any party to give us legal entitlement to the land, though every politician who visits villages like ours promises to do so,' says Tai. 'We're not a priority. We understand that.'

In rural India, land ownership is inextricably linked to social and caste hierarchies, and the landless are invariably those who face other forms of marginalization and historical injustice. Marathwada's *gairan* cultivation, now undertaken by the third generation of the one lakh Dalit families who occupied the land from the 1970s, has its roots in the pre-Independence era, when Dr Ambedkar had demanded from the Nizam of Hyderabad that waste or fallow lands be distributed to the landless castes to assure them of a means of livelihood.

Vishwanath Todkar, veteran land rights activist and secretary to the Jameen Adhikar Aandolan (land rights movement), a movement spanning almost twenty-five organizations across eight districts and covering hundreds of villages that have been agitating on the issue since 2000, says the first government order in favour of the Dalits came from the Nizam, around the time of Independence. Then too, continuing a historical tradition, upper castes opposed the transfer of land, until the *bhumiheen* (landless) satyagrahas led by Dadasaheb Gaikwad and others emboldened marginalized communities to start cultivating government-owned lands.

All eight districts in Marathwada have a higher percentage of population belonging to the scheduled castes than the state total of 11.8 per cent; in Latur and Nanded, this number is over 19 per cent, as per the 2011 census. This means the primacy of caste as a lived experience for the landless and assetless cannot be understated. For this sizeable number of people, landlessness is a stigma, but it is also central to their struggle for economic progress, especially amid recurring drought years. And if the *bhumiheen* satyagraha of 1959 underscored the significance of land reform and land redistribution, the Marathwada riots of 1978 offered evidence of sharpened caste-based atrocities against Dalits who were slowly making economic progress or engaging in socio-political mobilization.

The Marathwada riots of July–August 1978 began as soon as the Sharad Pawar government of the time announced the renaming of Aurangabad's Marathwada University to

A view of Jayakwadi Dam in Paithan, Aurangabad, the biggest in the region, in August 2019.

A Marathwada meal. Moringa or drumstick curry pops up in lunch unfailingly. The moringa grows fuss-free in drought-hit backyards, requiring only sparse watering.

A woman at work in a settlement of the Kolam tribe in Kinwat, Nanded. Many families subsist on traditional occupations such as basket weaving using bamboo and other materials from the forest.

Former Chief Minister of Maharashtra Ashok Chavan, who belongs to Nanded, at a Congress party event in Ausa, Latur, demanding government action on the drought conditions prevailing in Marathwada in May 2016. (*Photo courtesy of the Maharashtra Pradesh Congress Committee*)

Then minister Pankaja Munde at a '*shram daan*' (voluntary labour) event in Radi Tanda, a hamlet in Beed's Ambejogai taluka, in May 2016. The women were labourers at a road construction site. (*Photo courtesy of the BJP unit in Beed*)

As a political event progresses in Latur, rally-goers seek out the shade of a tree.

Villagers of Chaklamba, Beed, at a relay fast in June 2019 to demand a sub-canal from Jayakwadi Dam for their region.

Sweet lime trees reduced to charred skeletons due to water scarcity in Rohilagad, Jalna, in March 2019. Farmers have suffered repeated losses on account of drought, often spending lakhs of rupees in summer to fetch water through tankers.

Usha Shinde and Suman Raut of the Prerana Self-Help Group in Ramnagar, Jalna, enjoying a laugh while making lunch on a hot June day in 2019. Other than financial security, successful SHGs also often result in kinships that become a source of support for women.

Deepak and Kaveri Nagargoje at Shantivan in Beed's Shirur Kasar taluka. Shantivan is a school and residential facility for the children of migrant workers and tragedy-hit farmer families in the region.

Datta Bargaje and his wife Sandhya at Infant India, the home they run for the region's HIV-positive children who need shelter and care, outside Beed town. Infant India is located on a hillock opposite Bendsura Dam.

An aerial view of a cattle camp in Mhaswad in Western Maharashtra. It was launched on 1 January 2019, when the monsoon was still well over six months away. The camp, operated by the Mann Deshi Foundation, sheltered some cattle owners from the drought-hit parts of Beed in Marathwada, more than 170 km away. (*Photo courtesy of the Mann Deshi Foundation*)

Jalna farmer Balasaheb Dake on his farmland in mid-June 2020. Rainwater inundated his just-sown fields as large parts of Marathwada received unexpectedly heavy showers at the onset of monsoon. After losing multiple crop seasons since 2012 to acute water shortage, the region witnessed repeated bouts of excess rain and floods in 2020. (*Photo courtesy of Balasaheb Dake*)

Trees planted in Kinwat, Nanded, under a greening effort by the state government in 2018 were bone-dry the next year.

Tankers refill from dead storage in Bendsura Dam, Beed, in April 2019. Dead storage refers to water levels lower than the dam's spillways and outlets, requiring water to be pumped out.

Dr Babasaheb Ambedkar University, a decision that followed a years-long campaign by Dalit groups. What started as strikes and calls to boycott colleges in the big cities of Marathwada rapidly turned into organized, vicious attacks on Dalit colonies, homes and farms in rural areas. Sporadic incidents of violence against Dalits continued in rural Marathwada for about a month.

After spending twelve days visiting fifty riot-affected villages in Nanded, Aurangabad, Beed and Parbhani, a twenty-five-member Atyachar Virodhi Samiti, or a committee against oppression, comprising trade unionists, students, teachers and activists wrote a scathing report later that year about the violence, detailing instances of rape, molestation, arson, slaughtered livestock, polluting of wells with poisonous chemicals, assaults and murders of Dalits. Additionally, this citizens' inquiry panel observed that one peculiarity of the agitation was the 'direct participation of landlords', who organized meetings and led processions. The atrocities had been the gravest where some land reforms had taken effect, or where Dalits had organized themselves. Some villages had earlier witnessed reforms including permitting Dalits to draw water from village wells; in others, the surplus land had been distributed to landless Dalit families. The team wrote that *savarna*, or upper-caste, landlords clearly resented the Dalits' progress. 'In fact, atrocities also took place where the Dalits had somewhat improved their standards of living,' the Atyachar Virodhi Samiti wrote. The caste system was continuing to function as a tool of economic exploitation.

From 2000 onwards, the Jameen Adhikar Aandolan led by former Dalit Panther Eknath Awad began to demand implementation of a 1991 government resolution of the Sharad Pawar government. That resolution had stated that *gairandharaks* (possessors of *gairan* land) who have proof of having cultivated such land since before 1991 would be given legal title to the land. The resolution was drafted and published but never implemented. From 2000 onwards, at least 50,000 families submitted claims over such land in about 2,000 villages across Marathwada. Each claim was submitted, with the required documents, to six different offices – the chief minister, the revenue minister, the district collector, the deputy collector, the tehsildar and the local police station. Alongside, the Jameen Adhikar Aandolan staged protests and led rallies to the offices of the state administration. In 2008, at a meeting of *gairan* cultivators in Parli in Beed district, 3,000 people gathered – mostly from the Mahaar and Maang communities but also some from tribal communities.

Because the implementation of the 1991 government resolution is still anticipated, every single *gairandharak* continues to be particular about maintaining records as evidence of his occupation of public land. Every few years, tillers approach local police and revenue department officials and concede that their cultivation is an encroachment on government land. They request for penalties to be imposed, which they pay. The receipts, evidence of their continuing encroachment on the land, are carefully filed away.

The NGO Paryay, established by Todkar, and the Jameen Adhikar Aandolan have now prepared files on scores of villages where Dalits occupy *gairan* lands. Every file lists the number of such occupiers in that village and contains proof of each individual having been cultivating that land since before 1991. The file for Lakhangaon lists Tai's land, with letters about penalties from the 1980s. There are also photographs of the cultivators standing amid their crop. Tai and her family attended protests in Aurangabad and in New Delhi in the early part of the decade, and she knows that the issue has been placed before every government since the 1990s. 'But governments change, the new people [in power] don't know the history of our struggle, and it is back to square one for us. But the land is in our possession, and the officials now no longer try to drive us out of here.'

In Para village, which is just five kilometres away from Tai's home in Lakhangaon, Vijay Shingane, forty-seven, says the occupation of *gairan* land began in his village in the early 1980s, when Dalit families began to cultivate two to four acres each. Some distance away, a hamlet of landless people belonging to the Pardhi community, a denotified tribe, also occupied land simultaneously. About thirteen Pardhi families cultivate nearly thirty acres of land in a four-kilometre radius of Para. Shingane makes sure there is a '*panchnama*' or official assessment of his land in the presence of witnesses every year, following which he pays a penalty for the encroachment.

The recent back-to-back drought years have led many families with *gairan* land to also look for work as labourers, or to start trading in fruit or other farm goods. Many have had to try their luck in a big city. As the land is not legally theirs, they cannot avail of crop loans or other government schemes for agriculturists. They are also ineligible for government aid or compensation for crop damage due to drought or other climate disturbances. The land was to have been, at least partly, a ticket out of deprivation, but in recent years they have had one too many failed crop seasons. 'But we will not leave the land. We will not leave our struggle midway,' says Shingane.

For cultivators like him, life revolves around this government-owned land in their possession. Where they were earlier unwilling to sink money into a borewell or other farm infrastructure out of fear that the local authorities would object, those fears have now abated as district and taluka officials become more sympathetic to their cause.

In Para, a group including Shingane is planning to pool in money for a well – they are just not sure if that makes any sense right now, when officials have been saying that the groundwater table is at a record low. Still, the *gairan* cultivators invest their best resources in the land, for one bountiful harvest can set them free from daily wage labour for a few months. Additionally, the struggle has also given them community and a confraternity.

In 2009–10, this mobilization of the *gairandharaks* as a composite force began to set off new manifestations of caste-based discrimination and violence in some villages as a good crop season led to increased tensions with the village's power elite, especially as farm labour grew more difficult to find or became more expensive. In the nearby village of Gojwada, for instance, where Dalits lay claim to about fifty-five hectares of *gairan* land, there were complaints in November 2014 of their *tur* and jowar crops being destroyed by animals that had allegedly been driven into their fields by the village's upper-caste leadership. As tensions erupted, a house on *gairan* farmland was set on fire. Police complaints followed from both sides, and the mood in the village remained tense for months. In Para too, a case pertaining to caste-based violence was filed about fifteen years ago.

In Arajkheda village in neighbouring Latur district, where sixty-five acres of grazing land has been occupied, caste tensions between Dalits and upper-caste Marathas led to a social boycott in 2010–11. For over two years, Dalit families did not buy anything from local shops. They were also prohibited from working on the farms of Marathas. Here too, police cases were filed, including a complaint under the stringent Scheduled Caste & Scheduled Tribe (Prevention of Atrocities) Act, which only sharpened the fault lines and there was an undercurrent of anxiety in the village. During that period, the Dalits of the village said they had to travel a hundred kilometres to Latur city if they wanted to find work as daily wagers. There were other similar cases: a Dalit home was allegedly burnt in Latur's Neknaal village after *gairandharaks* tried to enter a disputed piece of grazing land. In one village in Aurangabad district, animals were driven into fields of standing crop in 2010.

Those days are behind them, say the cultivators. Even when tensions between Maratha and Dalit communities were at an all-time high in the state in 2016 and 2017 as Marathas pressed for reserved quotas for their community in government jobs and higher education institutions and dilution of the Atrocities Act, there was no direct conflict. 'It appears that those who earlier opposed the idea of us getting legal titles to this land are no longer interested in opposing it. But nobody is interested in taking our struggle to its logical conclusion either,' says Tai.

There may be relative peace and dignity for present-day *gairan* cultivators, but the long rural distress in Marathwada and the uncertainty about their land's legal status combine to prevent any social mobility, keeping caste hierarchy firmly in place. The way caste dominance operates in Marathwada's villages may have undergone a transformation, but the old power structures continue to dictate access to opportunity. Some *gairan* cultivators' sons in Mumbai or other big cities look for jobs that are not caste-ascribed, but they still sometimes find it difficult to blend in with the middle classes, often able to afford only inexpensive living quarters with no financial support from their families.

~

THE *GAIRAN* CULTIVATORS' ongoing struggle to gain land rights is a central narrative of Marathwada's deepest social inequities. But other disadvantaged groups have also been facing similar injustices for years. One example is the case of the Gond and Kolam tribes, who live in parts of Nanded district. In 2015,

Bhimrao Ramji Keram, then a Maharashtra legislator, wrote a flurry of letters including one to the National Commission for Scheduled Tribes. Keram represents Kinwat, a tribal-dominated constituency in Nanded, from where he was once again elected in October 2019 on a BJP ticket. Members of the Gond and Kolam tribes in Kinwat were being denied access to various special provisions made for them, he said in his 2015 complaint letter, because hundreds of non-tribals had been granted scheduled tribe certificates, thereby crowding out real tribals from institutions and government jobs.

In Maharashtra, the groups named Kolam, Mannervarlu and Koli Mahadev are recognized as scheduled tribes. However, in a fact conceded by the Maharashtra government's tribal development department and its Tribal Research and Training Institute (TRTI), several non-tribal people have taken advantage of similar-sounding caste or community names to claim that they belong to a scheduled tribe. In response to Keram's representation, the tribal development department wrote to the commission citing a report of the TRTI dated January 2016. The 'factual position', the letter said, was that taking advantage of 'similarity of nomenclature', there had been instances of people snatching benefits meant for scheduled tribe communities. For example, some members of communities such as the Koli, Sonkoli, Suryavanshi Koli and Mangela Kolis have claimed to be Koli Mahadevs. Some from caste groups such as Munurwar or Munarwar and Munarkapu claim to be Mannervarlus, a sub-group of the Kolam tribe. The state's caste scrutiny committees have come across several such cases, running into hundreds, in which the 'bogus' tribals have managed not only to obtain certificates naming them as Kolams but also secured government jobs reserved for scheduled tribe candidates in and around Kinwat.

Some officials say the problem dates back to the late 1980s, and there are cases of a second generation of non-tribals continuing to enjoy benefits reserved for tribal groups. There are several communities of genuine Kolams and Gonds, who often live together, spread across Kinwat. Though local officials and the tribal development department have admitted that there is a serious problem, state action to reverse it is slow and local groups continue to seek justice.

Similarly, barring some regions and brief periods where a certain district collector was sympathetic to the cause of the *gairan* occupiers, the 1978 and 1991 government resolutions that set out eligibility criteria and norms for regularizing the encroachments on *gairan* land were never implemented. Various individual cases went to court, where some still remain pending. But the Jameen Adhikar Aandolan's constituent groups dialled down their protests by 2014–15. For one, the land they had occupied was largely in their possession, a general awareness had spread that these encroached patches of land were the sole source of livelihood for the encroachers, and as wider agrarian distress began to take root in the region, priorities shifted imperceptibly among village- and taluka-level opponents of the *gairandharaks*.

'Also, as more people in government began to comprehend our long struggle, officials at various levels began to cooperate. District collectors and others assure us that we will not be removed from the land, that the land will remain in our possession, and that drafting a policy for legal entitlement to the land is not in their hands. That's why it made no sense any more to keep holding agitations and protest rallies at the office of the collector or *tehsildar*. Who should we protest to if the officials are supportive? There was no need for that any longer,' says Todkar. After 2014–15, the tightening of norms for NGOs' funding sources also meant constituent organizations of the Jameen Adhikar Aandolan were more selective about programmes.

In 2010, a public interest litigation was filed in the Bombay High Court seeking to regularize the encroachments on *gairan* lands as per the 1991 resolution. The case was disposed of in 2017, with the court observing that a blanket direction to the authorities to regularize the encroachments would not be possible. Meanwhile, in 2011, the state government issued another resolution, declaring that the encroachments on community-owned land meant for public purposes would have to be removed as per a Supreme Court order. That resolution too, however, was not implemented extensively.

Then in 2014, as the state government stepped up its social forestry programme, with targets for crores of trees to be planted, local officials who found there was not enough land to conduct these plantation drives received permission to begin afforestation activities on *gairan* lands. According to Todkar, the social forestry project planners had sought approval from district collectors to begin plantation only on unoccupied grazing lands. By the time this order trickled down to the villages, all grazing lands including cultivated *gairan* lands were legally transferred to the government for afforestation. 'But the lands are under occupation, and are being cultivated, and our claims for legal entitlement to the lands are pending. In some villages, officials tried to take over the land, but our organization stood united and firm. We submitted a fresh compilation of our records to Mantralaya,' he says, referring to the state government headquarters in Mumbai, 'and they backed off.' It was an indication that the decades-long struggle for land rights would now have to move to state-level advocacy.

In Lakhangaon, the cotton plants were still on the fields post the picking season in 2018 when officials arrived and began to take photographs of the land, apparently to begin a tree plantation drive. 'They also took away a pickaxe left on a person's farm and some rope meant for tying up bundles of the cotton plants' foliage,' says Tai Survase. 'When we heard about them walking around our land, the women rushed to the spot and chased them away. We told them either we would die today, or they would.' The officials left, but not without sowing a seed of worry among the *gairan* cultivators.

Then it was time to sow a rabi crop, and local tractor and bullock owners refused to rent out their machines or animals for the ploughing of the fields. '*Gairan* farmers do not own animals or machinery; we usually take these on hire. But nobody wants their equipment destroyed or seized by government officials, so we finally requested the sarpanch and the village's dispute-resolution team to be present themselves as we ploughed. The gram panchayat members gave us five pairs of bullocks for those days, and the sarpanch was standing right there on our land throughout the day while we worked,' says Tai. A former sarpanch herself from 2010 to 2015, Tai says support from the villagers to the *gairan* occupiers is critical in the absence of any active government action in favour of legalizing the encroachment.

'This was an irrelevant sequence of events that led to the land's legal possession being handed over to a government department for social forestry. Government officials themselves have earlier certified that the land is under cultivation. The least we expect now is a reversal of the order giving this land for afforestation,' says Todkar.

Not getting legal entitlement to the land has meant that through the difficult drought years, the *gairandharaks* have been unable to access government schemes and subsidies, compensation for crop loss, crop loans or crop insurance. In years of consecutive drought, where small farmers who are beneficiaries of the government schemes are forced to migrate in search of work, the landless and resourceless have been driven to the cities rapidly. The younger generation especially, say Tai and Todkar, is unable to wait any longer in uncertainty. At meetings or the occasional gathering of *gairan* cultivators, attendees are all aged forty-five and above. There is a sense that the land yields so little that even if it were to become theirs legally, the next generation's hopes cannot be pinned on it. Also, educated youngsters prefer to find service jobs instead of toiling on the fields for a few thousand rupees in profits each year.

'The previous generation,' says Todkar, 'was focused on the village, neighbours, status of the village, agriculture and so on. The new generation, however, is not so committed to the idea of land ownership for Dalits any longer. There are too many financial setbacks involved, so they see farming as an unsustainable occupation. And migration among them is very high.' After a struggle spanning nearly five decades, the paradigm of what constitutes a better life for the marginalized has itself undergone a transformation.

SEVEN

FLEEING HOME, FINDING HOMELESSNESS

~

'कामगार आहे मी, तळपती तलवार आहे
सारस्वतांनो! थोडासा गुन्हा करणार आहे'
(I'm a worker, a flashing sword / Saraswat pundits,
I'm about to commit some crimes.)

—Marathi poet Narayan Gangaram Surve

LAXMIBAI SHIVAJI JADHAV has tough, leathery palms from five years of working on construction sites in Mumbai. She treats the calluses by mostly ignoring them, occasionally willing them away. But they are hard to forget. There is a rough reminder every time she washes her face or mops her brow. Her face reddens where the calluses touch it. 'Maybe it's just as well that I can't pet my little boys every day,' she says dispiritedly.

Her sons, in Class V and VII, live 575 kilometres away in a private hostel, beyond the reach of her coarse hands. A web of intricate problems has led Laxmibai to live apart from her boys.

For five years now, the Jadhavs have lived seven to eight months of a year in a tiny home in suburban Mumbai. Nestled at the bottom of a hill filled with row upon row of semi-pucca shanties, their home in Bhatwadi, Ghatkopar, is so small that Laxmibai and her husband Shivaji must lay their sheets perpendicular to the door when preparing to turn in for the night. They stoop to enter, and most of their belongings are stacked outside. Every morning, Shivaji stands at his doorstep, towel wound around his wiry frame, brushing his teeth and spitting in the lane. On winter mornings, there is ash left behind from the little heaps of garbage they burn to stay warm through the night, and Shivaji aims for the ash as he spits.

By 8 a.m., the Jadhavs walk down the lane where other daily wagers gather as they wait to be picked up by labour contractors. The men gather first, bantering in Banjara, their native dialect. The women hurry out later, each one carrying lunch for two, often only *bhakri* and a chutney, sorghum or millet flatbread with a blend of green chillies, spices and crushed peanuts. Water is refilled into worn-out plastic bottles that once held mineral water. They meet their neighbours at the Bhatwadi junction, people they have known almost all their lives, all of them belonging to a cluster of five or six villages of Mukhed taluka in Nanded district.

The Jadhavs are from Manu Tanda, a hamlet where they own a little less than two acres of stony land atop a hill. Around the village lies a section of the Balaghat range of mountains, impervious and unsympathetic. The Jadhavs and other residents of Manu Tanda depend on wells and river streams for water. Laxmibai does not have a well on her land. About ten years ago, when she began to hear about a rash of new borewells being drilled in the neighbouring Latur district, Shivaji laughed. 'We're on top of a hill. Every farming season, when we go back, we have to start by removing stones from our land. Not pebbles, mind you, but large pieces of rock. If you hit my land with a pickaxe, you would likely hit a rock,' Shivaji says. 'Nobody digs borewells in Manu Tanda; it would just be silly.' *Tanda* is the Marathi word for a hamlet, mostly peopled by members of a single community, often backward or Dalit, and sometimes located just outside a larger village. The Banjaras' ancestors were originally nomadic, and they are categorized as an Other Backward Class.

Shivaji Jadhav, in his late forties, is nothing like the average migrant living in Mumbai: he's a graduate, a rare feat for the Banjaras. But like almost all the able-bodied men in his *tanda*, he is now a casual labourer. And like 50 per cent of the men in his hamlet, he is a Marathwada migrant in Mumbai. Manu Tanda and the villages around it, just like thousands of others in Marathwada, send men and women in droves to Mumbai every year, not all of them construction labourers, though that is the easiest unskilled job to find.

The state government has no definitive data on migration from drought-hit Marathwada, but in Bhatwadi, in the slums of the Kalwa–Mumbra belt, in the slum colonies engulfing the creekside and marshy areas of Mankhurd and Govandi, at construction sites across the Mumbai Metropolitan Region and at various fixed spots where the 'naka' workers gather, named so after the traffic junctions or *nakas* from where they are picked up in trucks and tempos by contractors, there are tens of thousands of those exiled by Marathwada's drought of opportunity.

In Bhatwadi, not more than 4 or 5 per cent of families have a child living with them. Like the Jadhavs, almost every couple has left behind some or all of their children at home with a grandparent or a relative, or have admitted them to one of the handful of private hostels that mushroomed in the larger villages and taluka towns of the region. The hostels cater to the specific need of Marathwada's farmers or landless workers who migrate to the big cities for employment, leaving home for a life of even more uncertainty, but determined to provide their children with a measure of stability and an education. Some decide whether to bring a child depending on the amenities they are able to arrange, such as daycare in the form of a neighbour, proximity to a municipal school and secure living quarters.

'If we don't come to Mumbai, we'll have to either die of hunger or kill ourselves,' says Kashinath Pawar, one of Bhatwadi's oldest Banjara refugees from Marathwada. When Pawar first arrived in Mumbai twenty-five years ago, Bhatwadi was an idyllic hillside in suburban Mumbai, and Bandra Kurla Complex, now a bustling commercial hub, was still a bog. 'Farming was always difficult in Abadi Nagar Tanda,' Pawar says of his native village, its name literally translated to mean 'populous city'. 'But in the last ten years, things have become worse there. Because even when there is one decent crop after two failed ones, the produce fails to fetch a decent price.' He thinks the pricing problem of agro-commodities is linked to the modern marketplace where shortfalls and gluts are no longer limited to a single region that a farmer can study and analyse. 'Now you're seeing even the onion farmers of Nashik taking their own lives. That never happened earlier; they were the somewhat bigger, more successful farmers,' he says without any emotion. As far as he is concerned, there isn't any hope for his lot, and the annual migration of young men from his village into Mumbai will continue unabated.

Among Bhatwadi's youngest migrants, Arvind Jadhav says he is nineteen years old, but he looks barely sixteen. He arrived in 2018 from Rampur Tanda in Udgir taluka, Latur. 'There were problems at home,' is all he will offer by way of explaining away why he gave up schooling midway. He is living with his sister, Shyamka Rathod, and her husband, who have been in

Mumbai for a couple of years. In the nine months that he has been in the financial capital, he has never watched a movie in a cinema hall, nor ever gone to sit by the sea. There's a 'Kabaddi Premier League' underway at a municipal ground right across the Bhatwadi hillock, but he has never attempted to visit. 'Where is the time?' he asks. He leaves Bhatwadi with the other Banjara workers around 8.30 a.m., returning by 7 p.m., with just enough time and energy to wash, eat, make a couple of telephone calls back home on his new Jio phone, play with his niece and then head to bed, tired to the bone.

IN SALGARA DEVTI village of Osmanabad's Tuljapur taluka, Savita and Eknath Lomte, both in their late thirties, see each other a few times every year. Eknath works as a casual labourer on contract at a fibre optic manufacturing company in Silvassa, Gujarat. The pay is about ₹400 per day for an eight-hour shift, all thirty days of the month. Leave comes with a loss of pay, so he takes a month off around every quarter to return to Salgara Devti, help at home, ensure the children's higher education is on track and to see if Savita needs anything specific. Three years back, when Savita was unwell with a serious gynaecological problem, she had to seek help from a neighbour. At the time, few in Salgara Devti had a latrine at home, and Savita needed help even to do the daily one-kilometre trek to the brushland that the women used as a toilet. They have a son and a daughter, both in college. About fifteen young men from Salgara Devti are employed as contract labourers in Silvassa, Eknath says.

In nearby Walwad, four to five young men found employment in 2018 in Alandi, a temple town near Pune, about 300 kilometres away. They work in the service and hospitality industry as helpers. In Beed's Patoda taluka, 80 per cent of

Bedarwadi village's people left home at the height of the 2016 drought, turning it into a ghost village where only very old people lived, including some unable to even fetch water from the tankers. These are only a tiny fraction of the desperate tales of those who leave, and of those left behind. And in each of these villages, residents point to the same frustrating set of circumstances.

Farming is simply not an occupation of choice when the income from it is so unreliable. The cost of cultivation has risen sharply; a bag of seeds, a sack of fertiliser, labour, power, drip irrigation systems, mulching sheets and pump sets have all become more dear, but the prices of agricultural produce have remained depressed. As yield per acre improves, some commodities run into a glut in some years. At other times, local produce competes with imported fare. Those with perishable commodities incur deep losses every time their delicate market mechanism is disrupted.

Right after the Union government demonetized the currency notes of ₹500 and ₹1,000 in November 2016, for example, cash-operated markets in mofussil towns simply wound up overnight. In Beed's Dharur taluka, without state-run cold storage facilities or any resources to afford refrigeration systems themselves, owners of nearly 50,000 milch animals found their home finances further wrecked after the kharif crop failed that year. Weeks later, ATMs were still not functioning. Even pensioners, who depend entirely on cash withdrawn from banks, were given a withdrawal ceiling at nationalized banks – there just wasn't enough cash. With currency notes sucked out of the market, poor farmers with milk or *khowa* to sell found they had to settle for half or one-third the usual rate if they wanted to be paid in cash. This class of very small farmers depends on daily cash income and were the worst hit, and that too during a drought year when bank loans and shop credit had already dried up. Many from these subsistence-farming communities are now annual migrants to Pune and Mumbai.

Meanwhile, government efforts to industrialize pockets of Marathwada to energize employment generation remain a non-starter. Under the Delhi–Mumbai Industrial Corridor that has been in gestation for a decade now, Aurangabad was to get two commercial nodes, Shendra and Bidkin. They were renamed the Aurangabad Industrial City or AURIC a couple of years ago and projected in official pamphlets as one of the country's best-planned industrial cities, occupying 10,000 acres and to be equipped with state-of-the-art infrastructure and technology. Work continues, but the proposed employment generation remains on paper.

The 2016–17 plight of the small milk producers of Dharur should have caused an upheaval for Marathwada's planners. That is because Dharur, a hilly, backward and drought-prone taluka, is seen by many locals as an ideal spot for the dairying business. Home to about 50,000 milch animals, its unorganized milk market is a bi-weekly bazaar held on Mondays and Fridays where purchasers including domestic consumers, small and large restaurants, caterers and others sample and buy fresh milk directly from farmers. On other days of the week, in the absence of cold storage facilities, the milk has to be turned into *khowa*, made from condensed milk, that is a staple in most Indian sweetmeats.

Milk producers either make the *khowa* themselves or sell to operators of *khowa bhattis*, stores where milk is thickened on stove-tops while being stirred continuously until all that's left is the flaky milk solids, or *khowa*. Because there is no market on other days of the week, and because neither the milk producers nor most of the *khowa* units own a refrigerator, the product is kept cool in clay pots hanging from roof beams. *Khowa* has a longer shelf life than raw milk, but only slightly. On very hot days, it splits, rendering the labour a waste. *Khowa* prices are seasonal, reaching ₹170 per kg in the festival season of August–November (2018) and otherwise hovering between ₹120 and ₹150 per kg.

When a young aspiring dairy entrepreneur conducted a survey in December 2017, he estimated that Dharur's daily milk collection was close to 2.6 lakh litres. He also calculated that about 1–1.5 tons of *khowa* is purchased from Dharur's weekly markets, which then travels to bigger towns such as Parbhani, Majalgaon, Pathri and even Pune. But the milk business provided only subsistence-level earnings for farmers, and the milk and *khowa* businesses generated negligible employment.

Dairying is very quickly turning into a go-to solution for agrarian communities struggling with cropping and crop pricing, and Beed would be an obvious hub in Marathwada, with its very large number of livestock. Right through a drought, thousands of families in regions such as Beed subsist on selling milk or *khowa*. Traders often purchase milk from the occupants of state-sponsored fodder camps, where the free fodder allows livestock owners to make small and life-saving profits through a period of drought. But milk from somewhat remote areas such as Dharur's villages cannot hope to reach larger markets until the infrastructure comes up, including a purchase network, cold storage and cold transportation facilities, and possibly a nearby dairy.

The aspiring dairy industrialist's first challenge was land: plots in the twenty-five-acre MIDC in Dharur were mostly taken already, even though not a single unit was operational. He could have got a job and moved to Pune or Mumbai but felt keenly that he must do something productive in and for Dharur, something that will bring success to many people. Until December 2019, he had not made much progress. This is despite the fact that any number of experts have pointed to the need to focus policymaking on dairying as an additional income generation avenue and also to generate employment.

The Kelkar Committee Report in 2013 said expressly that complementary businesses such as animal husbandry and milk processing, as well as making available veterinary experts, would greatly help the landless farmers in Marathwada. It also flagged fodder and milk production as areas for likely acceleration of sectoral and regional growth through a 'Fodder and Livestock Improvement Mission'. The Economic Survey of India for 2018–19 says as much: 'Livestock, poultry, dairying and fisheries is a sub-sector of agriculture that provides livelihood to agricultural households during phases of seasonal unemployment.' In discussing livelihoods, the central government's Manual for Drought Management notes that dairying and marketing of dairy products should be encouraged. But while India's dairying industry has grown steadily since the 1960s and 1970s, and the country now accounts for 20 per cent of the world's milk production, the experience in Marathwada is that there is no booster shot for smallholders in the sector.

With agriculture and agro-based industries both tottering, it is then a fairly straightforward decision, especially for the landless farmers, to leave the village. Vishwanath Todkar of the Jameen Adhikar Aandolan says the opportunities in the cities have also diversified for migrants who have had some education. 'A transformation is slowly taking place in Maharashtra's labour force of migrants. The largest numbers of the unskilled labourers in Maharashtra's cities are now from other backward states. Among the Dalits from Marathwada who migrate to Mumbai, many now manage to find slightly better occupations, working in shops or as drivers, for example, even though their living conditions in the slums is very poor,' says Todkar.

An activist with the Jameen Adhikar Aandolan for many years, Ashruba Gaikwad of Gojwada village in Osmanabad's Washi taluka is now a regular visitor to Mumbai, where his son has migrated. Gojwada is one of the hundreds of villages in Marathwada that witnessed the historic occupation of community lands meant for grazing animals in the 1960s and 1970s by landless Dalits. 'The younger generation is much less interested in fighting for a small parcel of land that mostly yields losses,' Ashruba says of the continuing struggle for land titles in Gojwada. His son and others from there live in the Govandi–Mankhurd slum belt of suburban Mumbai.

'Some of us have jobs at a store or a mall. Some have set up small enterprises, as contractors for electrical fittings or small civil works, or trading in something including agriculture produce,' says a young migrant from Gojwada. The youngster is not averse to farming, he says. 'But nobody is paying attention to the systems that need to be set right for small farmers to begin earning a decent living in Marathwada. Those circumstances must change first if the next generation is to take up farming.'

~

THE JADHAVS OF Manu Tanda, Mukhed, survived years of the same debilitating circumstances before they began their annual migration to Mumbai, where they live seven to eight months of the year, returning home to till their small patch for a few quintals of jowar that they then stock up for the family's use through the year. Their annual travel is mirrored by that of thousands of others who were born and have homesteads and families around the tourism and manufacturing hub of Aurangabad, but still travel nearly 500 kilometres to Mumbai or Pune to scrape together a subsistence lifestyle.

For the first time in their five years of living in Mumbai, Laxmibai brought her sons with her for the 2018 Diwali holiday – and regretted it right away. 'We were away at work all day, and the kids were just at home alone with nothing to do.' The boys still actually liked Mumbai, or whatever part of Ghatkopar they were able to explore themselves, catching a glimpse of the Versova–Andheri–Ghatkopar Metro Rail and the crowds at Ghatkopar railway station. Laxmibai and Shivaji managed to find time on four or five evenings during the boys' month-long holiday, and they visited a few streetside markets with them. But Laxmibai was relieved when it was time to send them back to hostel. In their five years in the city, the Jadhavs too have never visited a movie hall or a restaurant. They were never rich, but they had pride in their land back home. Laxmibai describes her home in Manu Tanda with moist eyes: it is in an uneven row of simple brick and cement homes with thatched roofs, but with wide-open spaces between homes.

Hanging on the arm of Bhatwadi newcomer Arvind is his niece Anita, all of four years, wearing a frock too long for her and a mischievous glint in her eye. Reluctantly, the family admits she has not yet gone to school. Her mother Shyamka, who gives her age as twenty-three, has an older son aged six, who is being looked after by a grandparent back home. Anita is still too young, says Shyamka, picking up her barefoot daughter. The girl lost her slippers several weeks ago at a construction site. Neighbours laugh as somebody passes around a mobile phone with photographs of the girl. In them, a grinning Anita is covered from head to toe in a thick layer of cement. 'Maybe she should be in school; it's not always possible to keep an eye on her while I'm working,' says the young mother. 'But who will drop her to school and bring her back and look after her during the day until I get back?'

Every Banjara home in Bhatwadi empties out in the morning, Monday through Sunday, everybody ready to work on public holidays, weekends and festivals too, if work is available. The few who have children at home simply must take them along to the site of their work for the day. Ironically, the lane where the labourers gather in the mornings before they are picked up by contractors leads to the Barve Nagar municipal school. Only a couple of years older than Anita, a row of children is herded in by teachers, plastic water bottles around their necks and weighed down by their satchels, all sleepy or glum, or both. Anita has no idea where they all march to every morning.

The problem of out-of-school children is rampant across Marathwada. The largescale seasonal migration, whether to Mumbai for casual labour or to sugarcane fields in Western Maharashtra or Karnataka, keeps thousands of children away from school for several months at a stretch, after a few years of which many just drop out. In 2009, alongside the enactment of the Right to Education law, mandating school education for all children, activists who work with migrant sugarcane harvest labourers prevailed upon the Maharashtra government to study the problems of these families' children. Consequently, in 2010, the Maharashtra government announced its intention to keep children of migrant agricultural labourers in school even when their parents migrate, and introduced the idea of seasonal residential hostels for such students. These were to be called the *hungami vastigruha*, or seasonal hostels, where children could not only attend classes but also reside for a period of six months in the year, with the state sponsoring the children's food and essential supplies through an appointed caretaker and cook.

These quickly ran into trouble – security standards, especially for girl students, the quality of food and the inevitable problem of corrupt contractors. Sudhakar Kshirsagar of Sankalp Manav

Vikas Sanstha, an organization based in the drought-hit Pathri taluka of Parbhani district, says the seasonal hostels were a good experiment, even if their impact is yet to be scientifically assessed. But with farm labourers travelling farther and farther, and others like the Mukhed residents being away for almost three quarters of the year, the government should consider extending the facility year-long, turning these into permanent hostels for the children of families suffering on account of distress migration. Kshirsagar's organization also conducted a study of sixteen existing seasonal hostels in 2014 and found lacunae ranging from the absence of security guards to lack of specific facilities for girls.

In any case, the Banjaras of Mukhed have never heard of these seasonal hostels. There are probably not enough of them, they say, and perhaps none in areas lacking a politically powerful class. The Jadhavs spend ₹25,000 per year per child to keep their sons in a private hostel, almost one fifth of their annual income. Those with marginally better means admit their children to hostels in Mukhed city, a tier III town about twenty kilometres from their village. Babu Jadhav, the sole Banjara in Bhatwadi to have purchased his own home in Mumbai after nearly fifteen years here, is one of them.

When he first arrived, he pitched a tent in the Barve Nagar municipal ground where the kabaddi tournament is now held. He lived there for the first two years, right alongside a large open drain, the odours wafting from it and his threadbare mattress constantly reminding him that he needed to move up the food chain. Having started as a labourer, Babu now takes on minor civil contracts for constructing or paving gullies in slums, public toilet blocks, repairing drain walls, and so on. He owns a smartphone and a motorcycle. Youngsters like Arvind call him from home before they arrive in Mumbai, enquiring if they are likely to find work. Of course, there is work, he tells everyone. There is plenty for everyone in Mumbai. He himself manages to employ ten to fifteen casual labourers on most days, paying them the standard industry wages – ₹500 per day per skilled or experienced male labourer, less if the worker is a fresher or completely unskilled, ₹250 or ₹300 for women, and ₹700 on average for a couple. 'Nobody here likes the work we do or the way we live in Mumbai. It's long hours of back-breaking work in the heat and dust; there are no benefits like holidays or festival bonuses. But at the same time, not a single person has ever returned home because they're unwilling to work here,' he says.

Towards the latter half of 2019, a few more youngsters from Mukhed arrived in Mumbai, bearing the familiar stories of repeated cycles of crop loss, bad credit, private credit and despair. He says, increasingly, the migrating workers are younger nowadays when they arrive, most of them just about eighteen. Some have completed Class X or XII and try to look for other work in place of hard labour. But for all, Babu says, even the squalor of makeshift shanties along fetid suburban Mumbai drains holds some value if it comes with a few thousand rupees sent home or saved each year.

Less than two years back, Babu purchased a small home in a row of squat chawl structures in Bhatwadi. He doesn't like calling himself a success story but admits that he learnt the ways of Mumbai and found his own little toehold. He is evidence of the possibility of returns from migration improving slowly, over time.

EIGHT

CROPPING A CONUNDRUM

~

'ढगाला उन्हाची केवढी झळ!'
(How the heat burns up the clouds!)

—From 'Aggobai Dhaggobai', a poem by Sandeep Khare in
Balbharati Class I Marathi textbook, overheard recited in a
classroom in Shirur Kasar, Beed

MACHHINDRA GAWADE OF Chaklamba village in Beed district's Georai taluka spends a lot of his time thinking and talking about water. Right through the five decades of his career as a socialist, political activist and farmer, Gawade Mama (uncle), as he is known, has been obsessed with water. As a young man, he lived in water-rich Western Maharashtra. 'So I have observed the politics of water from two sets of eyes, of those it enriches and of those it impoverishes by its absence,' says the seventy-one-year-old.

In 2018 and 2019, Gawade Mama and other Chaklamba villagers undertook a 'relay fast' multiple times to draw the attention of authorities to Chaklamba's unique problem.

Jayakwadi, Maharashtra's largest dam, across the Godavari, is just forty-odd kilometres from Chaklamba and eighty other villages in this cluster. From one bank of the dam, water gurgles softly through canals to fields in distant villages, including some located 400 kilometres away in Nanded district. On the opposite bank on the east, the canal skirts the Chaklamba cluster and cuts across the highway at a point just thirty-five kilometres north and heads towards Majalgaon.

Chaklamba and its surroundings received just over 300 millimetres of rain during 2018's south-west monsoon. In the windblown summer afternoons of June 2019, as they waited for respite after nearly ten months of acute water scarcity, the congregation of protestors near the Hanuman temple on the outskirts of the village was always large. The men had little or no work, most having had to skip the rabi season or winter crop entirely. Less than an hour's drive away, the foliage of the six-foot-tall sugarcane fields was lush, roots still moist. The east bank canal alongside was now dry, but the earth was still soft.

Gawade, Beed district chief of the Shetkari Sanghatana, a political party representing farmers' interests, has been petitioning the administration for a sub-canal from the east bank canal to the Chaklamba circle since 2016, with no success. Almost every well in Chaklamba and nearby villages is dry in the summer of 2019, and these villages depend heavily on government-funded or private tankers for drinking water. 'Our farms are barren; we don't have a crop. And forty to fifty kilometres away, farmers were flooding their cane fields with water until a few months back,' Gawade says bitterly.

Cane, labelled a water-guzzling crop by the Economic Survey of India, is a hotly contested subject in Marathwada. There is a growing chorus of outrage from environmentalists, water-sector experts and economists over the rapid growth of the sugar industry in water-scarce Marathwada, based on the premise that cane fields are sucking out from dams, canals and groundwater a disproportionately – unconscionably – large quantity of moisture. In dryland areas, this must surely come at the expense of other, poorer farmers who do not have access to dams, canals, lift irrigation schemes and the like, they believe. Cane growers and most farmers, on the other hand, are suspicious of any proposed interference in their right to select a crop of their choice. In fact, not even Chaklamba's summer-weary farmers take issue with sugarcane. They would opt for cane too, if they had an assured source of irrigation. But juxtaposed against the land that farmers are forced to keep fallow, the cane fields a few kilometres away are emblematic of wider debates over water-intensive crops in dryland areas. The startling inequity is quite plainly visible between subsistence farmers dependent on truant rains and others in the same region accessing irrigation facilities for highly lucrative but water-intensive cash crops.

Marathwada has always received lower rainfall than other regions of the state, with an average seasonal precipitation of only 779 millimetres. Over the past ten years, however, the average seasonal rainfall has dipped further. Between 2009–10 and 2018–19, average rainfall during the four-month south-west monsoon season was 684 millimetres. In 2012–13, average monsoon precipitation was 69 per cent of the region's long period average, dipping further in 2014–15 to 53 per cent and in 2015–16 to 56 per cent. In 2018–19, rainfall averaged 64 per cent of the long mean. During these drought years, seasonal rainfall was in the range of 415 to 540 millimetres, about as much rain as over three heavy-rain days in Mumbai.

In much the same period, between 2010–11 and 2018–19, the area under sugarcane farming in Marathwada, the number of sugar factories, the total crushing capacity installed in these factories, the actual quantity of cane crushed, sugar production and alcohol production in sugarcane factories all rose dramatically. The rising trend broke for two years in 2015 and 2016 as cane production dipped following poor rains, but the numbers still tell a lucid story of how the cane industry has come to bear rich fruit in Marathwada since 2010.

According to data from the state government, functional sugar factories including cooperative and private factories in Marathwada's eight districts numbered forty-six in 2010–11. In 2018–19, this number rose to fifty-four. The cane crushing capacity in these factories grew from 94,550 metric ton per day in 2010–11 to 1,57,050 ton per day in 2018–19. Actual crushing of cane grew from 130.03 lakh metric ton in 2010–11 to 194.31 lakh metric ton in 2018–19, or by about 50 per cent. Sugar production grew from 14.23 lakh metric ton to 20.91 lakh metric ton. The production of alcohol in these sugarcane factories grew from 579.86 lakh litres to 1,110.98 lakh litres. Many factors play a part in industrial production success stories, including better equipment or less factory wastage, and data is not immediately available on how much more water cane fields and sugar factories consumed in comparison to the previous decade, but it is safe to say such magnificent growth of cane acreage would imply more water consumed by the crop.

Anxious as the 2014 drought showed evidence of rapidly depleting groundwater levels, the state government feebly suggested demarcating areas as cane-permissible or otherwise. To many, it seemed the government was only making the right noises during a crippling drought year. The South Asia Network on Dams, Rivers and People (SANDRP), which has worked extensively on the issue of cane and equitable water distribution, said in a blog post in 2015: 'Maharashtra Government made an announcement in August (2015) that it may consider banning sugarcane crushing in Marathwada to protect drinking water supplies. This announcement made by Revenue Minister Eknath Khadse was met with uproar as well as scepticism … Expectedly, the announcement turned out to be a trial balloon and was taken back. Then the Cooperatives Minister Chandrakant Patil announced that crushing will commence, but the final decision whether to allow or disallow crushing will be taken by the District Collector of a particular district after analyzing existing water storages, water availability of the region till July and the impact that water diverted for crushing will have on the water security of the district.'

As the 2015 cane crushing season began, SANDRP compiled a report based on discussions with district collectors – even those officials who admitted that the water crisis was too grave to permit use by sugar factories said they had received no instructions from the state government on what steps to take. The crushing season progressed, though considerably dampened by the previous season's acute water scarcity. A decline in area and productivity of cane owing to the drought led to a dip in cane production in 2015–16 and a more significant decline in 2016–17, but there was no state measure to reflect the government's professed seriousness towards disincentivizing growing cane.

From the large biomass that cane accumulates as it grows, more than any other crop, it is visibly water-intensive. Cane growers know that maintaining proper soil moisture is central to a good crop, and waterlogged cane fields are the norm. But just how water-intensive is the sugarcane crop? Why does one crop elicit such resentment? A study in 2013 by the Indian Sugar Mills Association said cane requires 1,500–2,000 millimetres of water per year for every hundred tons of cane, or 150–200 lakh litres per year per hectare. Officials in Maharashtra say the average water requirement for the cane crop in Marathwada is about 200 lakh litres per hectare per year. Multiply that by the area under cane production, which is 3.13 lakh hectares in Marathwada (2018), and we realize the crop consumed 6,260 million cubic metres of water in a year, in large part from canals and dams and wells, because the monsoon was poor. That translates to a little over twice the capacity of the enormous Jayakwadi Dam at full storage level of 2,909 million cubic metres. And this does not compute water requirements of sugar factories for crushing operations, electricity, ethanol generation, etc.

A 2018 report by the National Bank for Agriculture and Rural Development (NABARD) and the Indian Council for Research in International Economic Relations (ICRIER) estimated that about 80 per cent of the irrigation requirements of sugarcane across the country are met through groundwater sources. It said the crop needs 'frequent light irrigations, each of 40 to 50 millimetres, adjusted to suit the growing period of the crop and to the prevalent weather conditions'. Water requirement ranges between 37.5 centimetres (Bihar) to 297 centimetres (Tamil Nadu) in about five to forty irrigations of 7.5 centimetres each per hectare during crop growth, the report said.

In Maharashtra, the requirement is between 206.3 centimetres and 243.8 centimetres during a crop season, it said, quoting a 2015–16 report by the Commission for Agricultural Costs and Prices (CACP). 'Sugarcane is one of the most water-intensive crops. This is evident from the fact that on average more than 96 per cent of the land under sugarcane [production] is irrigated and in many states like Maharashtra, Karnataka, Tamil Nadu, Haryana and Madhya Pradesh irrigation coverage is 100 per cent or approaching it,' the report said. This means in Maharashtra, where irrigation coverage is abysmal, hovering around 23 per cent of total cultivated land, one crop is sufficiently powerful so as to command 100 per cent irrigation.

The CACP's 2014–15 'Price Policy for Sugarcane' report said the area under sugarcane use in Maharashtra accounts for only 9.4 per cent of the gross cultivated area, but the crop appropriates 71.4 per cent of all irrigation water. This does not account for further water-intensive processes in the factories. In 2015, SANDRP found at least one factory in Marathwada cleared to draw 45 lakh litres per day from the nearby dam reservoir.

The 10,000-odd people of Chaklamba have no inkling of such troubling math. Seven thousand of Chaklamba's residents migrate annually to work as labourers during the sugarcane harvest season. 'Poor people from all communities have now begun to sign up with contractors for the cane harvest. Some travel into neighbouring districts in Western Maharashtra and some go longer distances into Karnataka. When they're gone, local shops and businesses come to a standstill, because there's no work and no income available, and anyway there are mostly only senior citizens and grandkids left behind,' says Krishna Khedkar, a resident.

Some who own six to eight acres of land in and around Chaklamba work as farm labourers on smaller fields in Western Maharashtra districts, because eking out a living from a crop on their own farmland is now no longer viable. Naturally, they have no quarrel with cane growers. If they ever get the sub-canal they have been clamouring for, many will try to become cane growers themselves.

~

AT SIXTY, BHAIRAVNATH Bhavanrao Thombre has had almost three decades to wrestle with the knotty ethics of cane. Chairman and managing director of Natural Sugars & Allied Industries, which has two factories in Marathwada's Osmanabad district and one in Pusad of Vidarbha's Yavatmal district, Thombre's background, from being one of four sons of an impoverished farmer to promoting and running large sugar factories, associated alcohol and ethanol units, and also schools and colleges in Osmanabad, sets him apart from many other sugar barons. He is no stranger to cyclical drought and loss of livelihood. When the withdrawing monsoon of 2019 – the meteorological department called it post-monsoon rains – came belting down during three weeks of October and November, Thombre was exultant. 'The rains washed away three years of drought and pain. And hopefully also washed away the deliberate, malicious opinions about the cane crop's impact on water resources. It has been so tiring to keep listening to so-called experts demand that cane be banned in our region,' he says.

Thombre, also president of the Western India Sugar Mills Association (WISMA), is one of the founding fathers of the cane industry in Marathwada, having worked with late BJP leader Gopinath Munde's Vaijnath factory in Beed and then with late Congress leader Vilasrao Deshmukh's Manjara factory in Latur before launching his own enterprise in 2000. Thombre was awarded the state government's prestigious Krishiratna award in 2018 as a commendation for the many water-saving measures implemented by Natural Sugars. The livelihoods created by the entry of his sugar factories and then a ferrous alloy unit in a backward region such as Ranjani village in Osmanabad's Kalamb taluka are the backbone of the village economy, he says.

Approximately 263 villages within a fifty-kilometre radius of the factory produce high-quality cane, but being a high-yield region there was clearly an excess of it, leading to depressed demand in the late 1990s. The Natural Sugars factory, set up in the 100 per cent irrigated area of Manjara Dam and in close proximity to the Raigavhan Dam, a medium-size irrigation project, works with cane growers who are able to produce a crop even in drought-like conditions. Over the last two decades, Natural Sugars has set up a distillery, a co-generation plant, a biogas plant, a bio-compost unit, a dairy and a sugar refinery. In addition, the company has set up a school and college in Ranjani, various water-conservation and sanitation projects, a medical facility and other member–farmer welfare schemes including one for pension.

'In the last fifty years, many big sugar factories have similarly expanded their activities. Together, these factories employ lakhs of villagers who otherwise have no job opportunities. The cane growers are shareholder members in cooperative factories; they are able to build assets and their children are able to access education,' says Thombre. 'I say let's get a clear idea of the economic development of backward and water-scarce regions on account of the sugar industry before we start blaming sugarcane for the drought.'

The contentious debate over watering sugarcane fields is not unique to Maharashtra alone, he points out, even if it is particularly intense in water-scarce Marathwada. Across India, sugarcane fields occupy just about 3 per cent of total cultivated area. And yet, cane is one of India's top five crops based on economic value. According to a 2018 report by NABARD, it accounts for 7.5 per cent of the gross value of agricultural production in India.

Thombre says where other crops have failed to provide returns, cane has been a reliable option. 'Ninety per cent of soya bean across Marathwada was destroyed by the recent untimely rains, and what was left is not fetching prices above ₹3,000 a quintal. With cane, there are no price depressions. As for how much water the cane crop needs, I invite agricultural universities to set up meters to generate precise data on the matter. Cane is not watered when it is raining, just like other crops. We must calculate how much watering is done for a cane crop that lasts about twelve months, and compare that with any other twelve-month crop such as a horticultural one. Or compare it with two back-to-back crops that together account for twelve months in the soil. I am confident that cane is not the water villain it is made out to be,' he says.

Natural Sugars is attempting to quantify this precisely, and early experiments have shown that a one-hectare field of cane uses about 150 lakh litres of water during one season, not 200 lakh litres as the experts estimate. With 100 per cent drip irrigation, the requirement for a hectare of cane can be further reduced to 75 lakh litres, Thombre says, though he concedes that neither has micro-irrigation been adequately absorbed by cane growers nor have other factories upgraded to water-saving systems.

The wider argument for cane is actually made by the farmers. A sturdy crop that withstands unseasonal rains, minor hailstorms and is not prone to pest attacks, it is a safe choice for farmers with assured water supply. In addition, factories are mandated by the government to pay a pre-fixed 'fair and remunerative price' or FRP to cane growers, in comparison to other cash-crop producers who face price fluctuations even during a single season, compete with imports, joust with export restrictions or have to arrange storage for their produce as they cool their heels waiting for prices to improve. Cane growers are immune to those complications, there is price and income stability, though the issue of outstanding payments or cane arrears from factories is a convoluted one, and factories sometimes go bankrupt. 'It is literally the most reliable crop from the farmer's point of view,' says Thombre.

But his calculations are rejected by multiple experts and official reports that have since the turn of the century warned that this outlook is myopic in view of choices to be made for long-term sustainability. These voices have grown more and more urgent in recent years. In 2007, before the string of drought years in the state, an audit report by the Comptroller and Auditor General (CAG) mentioned in passing that permissions for new sugar factories do not appear to consider water availability. The report said, 'As water is a critical input for the production of sugar, the Commissioner also has to scrutinize the project proposals of the cooperative sugar factories against the Ground Water Survey and Development Agency's certification regarding water availability in the area before recommending their registrations.'

It said some cooperative sugar factories had undertaken construction, maintenance and operation of irrigation systems for their use in order to shore up water availability, both for the cane growers of their region and for the factory's operations. At least two factories that the audit team checked showed evidence of 'injudicious' investment in irrigation or water supply schemes. One factory in Sangli had invested in storage tanks and lift irrigation schemes, but had incurred losses because members did not pay the imposed water charges. Another factory in Buldhana had planned to build an underground reservoir, pipe house, jack well and pump sets to lift water from the Purna River, but the work had remained incomplete though huge expenses had been booked.

In 2011, the Maharashtra Water Resource Regulatory Authority (MWRRA), the apex body to regulate the water sector in the state and to ensure judicious and equitable water management, recommended that total sugarcane crop acreage should not exceed 10 per cent of the command area of any dam. This norm is not followed anywhere. In fact, in the months after the drought of 2012, the sugar commissionerate in the state processed permissions for several new sugar factories in the districts of Beed, Osmanabad, Aurangabad and Jalna, much to the shock of environmentalists, prompting a further increase in cane acreage in these districts.

The extensively researched Kelkar Committee Report of 2013 on the backlog of development and regional imbalances in Maharashtra also noted that the popularity of sugarcane as a cash crop has led to 'extreme inter-crop imbalance in the consumption of water resources'. The report advised the state to involve these sugar cooperative factories to popularize micro-irrigation. 'The cooperative factories are primary users of sugarcane and have contractual liability of processing sugarcane grown by members. The significant portion of capital needed for establishment of factories has been contributed by the state government and the state has provided guarantee to the debt capital. Moreover, in view of persistent relative scarcity of water, government should regulate the use of water through appropriate economic instruments. As a policy, sugar factories may be incentivized as well as persuaded to support adoption of micro-irrigation techniques for their members,' it said.

It further suggested mandating drip systems. For each hectare of cane with drip irrigation systems, enough water can be saved for twelve hectares of pigeon pea. So, the virtues would be amplified, reflected not only in water conservation but also in the productivity of other crops. Dr Kelkar's other recommendations included cash support to farmer groups for water harvesting and conservation; and significantly, a 'limit on area under sugarcane in a given river basin as suggested by Godbole Committee and Water Commission (1999)'. None of this was implemented.

Multiple other reports have made near-identical points. The CACP's report on pricing for sugar in 2017 said that as water scarcity grows, there is a growing need to improve 'crop water productivity' so that there is optimal use of the limited freshwater resources. During field visits, the commission had noticed in other states that the traditional flood-irrigation was still most popular. In 2015, the CACP said that given that Maharashtra and Karnataka had experienced severe droughts in the previous year, it would be essential to optimize every unit of land and every unit of water. 'Against this backdrop, the Commission recommends taking up drip irrigation and fertigation on a much higher priority in drought-prone belts of Maharashtra and Karnataka, which has the potential to save almost 40 to 50 per cent water ...'

Maharashtra's former chief minister Prithviraj Chavan, who belongs to Western Maharashtra, said in an interview to the *Indian Express* in 2015 that more sugar milling in Maharashtra appeared unsustainable: 'I agree with water management expert Madhav Chitale that sugar mills should not be allowed in drought-prone Marathwada.' While relocating existing mills could be politically near-impossible, these mills too should be subjected to stricter norms on water use, he said.

Barring the powerful sugar factory lobby, there is a clearly emerging consensus that restoring equitable access to water through crop management is prudent. But, as water scarcity deepens, bold policymaking appears even more challenging. Because, paradoxically, the harsher the socio-economic impact of a drought, the more anxious poor farmers are to try and switch to a crop that promises a stable and secure income. Chaklamba's farmers have for the past decade watched their assets erode as they sold animals or parcels of land or gold. An assured source of water supply could help them recoup years of losses if they can coax out of their fields one season of sugarcane.

~

LATE IN THE summer of 2017, Girdhar Solunke of Nagapur village in Parbhani, a cane farmer for many years, decided it was time to buy a four-wheeler, not a tractor or pick-up truck but a car, maybe a sedan, something his family could use. In his early thirties, Girdhar owns only two acres of land, but the family's cultivation is done jointly, and their eighteen to twenty acres of sugarcane bring sweet rewards for their toil. Having decided to take an automobile loan, he filled out application forms and registered to get a CIBIL credit score. The score was low, and the loan could not be processed. The CIBIL report said he had an outstanding loan at the Nagpur branch of Andhra Bank.

'I had never even visited Nagpur. At first it seemed an obvious error – somebody had mixed up Nagapur and Nagpur, so my name as loanee was a mix-up too,' says Girdhar. 'Or so I thought.'

He boarded a train to Nagpur the next week and went straight to Andhra Bank. There was no mix-up – there was an outstanding loan of ₹72,000 against his name. The loan account bore his name, his photograph and accurate details of his other identifying documents such as his Aadhaar card and his agricultural land records. His heart racing, Girdhar pressed the bank staff for more information, and found there were similar loan accounts in the names of his immediate family members, including his father, uncle and cousins. Meanwhile, a friend attempting to obtain a loan from Uco Bank's Gangakhed branch was similarly rejected. Gangakhed is a taluka town in Parbhani, adjoining Beed district's Parli Assembly constituency. Over the next few days, it emerged that dozens of others from his village and nearby villages, all cane growers, owed money to Andhra Bank's Nagpur branch.

Within weeks, the story unfolded. Gangakhed Sugar & Energy Ltd, the sugar factory where Girdhar and others in the belt sold their cane, had allegedly used their documentation to obtain loans worth crores of rupees. They were individual crop loans, each one with a corporate guarantor – the Gangakhed sugar factory. In July 2017, Girdhar became the first complainant in the case, and Parbhani's biggest sugar baron, Ratnakar Gutte, the main accused.

According to court documents, the main allegation is that the company's representatives and others prepared false and fabricated documents to secure loans from various banks in the names of 'thousands of agriculturists'. Unknown to these farmers, the sugar factory on which they pin their annual hopes had taken crop loans under their names, utilizing the sum for its own benefit. As investigations proceeded, Gutte, known for his political clout, who was also a member of a party that was an ally of the government at the time, was not arrested.

Then in March 2019, a charge sheet was filed and with just weeks to go for the general election to the Lok Sabha, Gutte was arrested. One of the arguments he would subsequently make while seeking bail was that the charges had been brought upon him by political rivals. In opposing his bail, the state would tell the court that fictitious loan proposals were prepared and amounts secured in the names of nearly 24,000 farmers, sums totalling several hundred crores of rupees. Gutte stayed in jail, his bail plea rejected, but in October 2019, he won the election to the Maharashtra Legislative Assembly from the Gangakhed constituency, apparently running his campaign from jail.

At least some poor farmers on whose names Gangakhed Sugar took loans must have voted for him, shrugs Girdhar. 'That is just how relationships around here unfold,' says Govind Yadav, a political activist in Gangakhed. When the police began to investigate, they found that many farmers were not aware of loans having been obtained in their name. Several farmers who later approached the factory were able to get the loans against their names repaid, in full. Many simply want the factory to run; one less factory in a cane-rich belt could upset demand for their produce. Some farmers may be unconcerned about a poor CIBIL score so long as the cane grower–factory relationship functions smoothly. Further muddling the equation are the banks, which actually want to loan to small sugarcane farmers who are in general less likely than other small farmers to have a history of bad credit.

In the Gangakhed case, the factory was guaranteeing the loans to small farmers. So, the long-term issue around cane is not just sustainability and water use, or jeopardizing water security for future generations. To people such as Girdhar and Govind, the inequities kindled by cane are just as problematic, along with its apparent systematization of egregious power relationships. Sugar cooperatives and sugar barons enjoy political patronage of lakhs of member farmers who are a sizeable vote bank, but that patronage also has deeply feudal innards. And it is no longer a secret that the creation of irrigation potential and dams was often concentrated in cane-growing regions.

Meanwhile, the 2018–19 season saw 3.13 lakh hectares of farmland in Marathwada under sugarcane, up from the previous ten years' average of 2.16 lakh hectares. Cane now occupies 5.74 per cent of Marathwada's cultivable area and 27.97 per cent of its irrigated cultivable area. Despite having a much lower per-acre productivity in Marathwada than in the rest of the state, it grows more and more popular among farmers. Average cane productivity per hectare in Maharashtra is 85.37 metric ton, while in comparison it is only 57.28 metric ton per hectare in Marathwada.

In a less direct way, the question of asymmetrical access to water, which is central to the subject of the suitability of cane, is also enmeshed in the wider questions surrounding the logic of big dams. Shahdev Ashruba is among several villagers in Chaklamba who lost their land in 2000 when the state administration built a thirteen-acre water tank just outside the village. A simple embankment wall structure dammed rain streams and water from the Sindphana River rushing down the surrounding hillslopes. The very next year, the wall collapsed and was never rebuilt, even as the Maharashtra government stepped up spending on construction of dam projects worth thousands of crores. Shahdev, who owned just one and a half acres of farmland, lost nearly half of it to the tank project. The ₹16,000 compensation for his land in 2000 ran out before long, and he has since then had to additionally work as a labourer to supplement his meagre earnings from the farm. Villagers say the tank is such a simple and low-cost structure that there is no enthusiasm in any level of government to rebuild it, even though it would resolve at least some of Chaklamba's farm woes. Discussions about demanding that the tank be repaired sometimes grow very agitated and urgent, but further action eventually ebbs and flows, because getting a toehold in the system of political patronage that guarantees cane growers bonuses and state-administered prices seems to be a better idea overall.

The latest effort to warn the state government of a water crisis around cane was in August 2019, when a top official in the Aurangabad division, the headquarters for Marathwada's administration, wrote to then Chief Minister Devendra Fadnavis suggesting that no new cane-crushing licences be given unless factories' exclusive cane zones convert entirely to drip irrigation. A total ban on sugarcane in water-deficit sub-basins was also suggested. According to very senior bureaucrats, the report said Marathwada had been found by climate experts to be prone to climate change impacts. But having come just before the October elections to the Maharashtra Legislative Assembly, the report was mostly buried, except when vote-seeking politicians addressed cane-growing regions and promised to exorcise the report entirely, and never implemented a word of its recommendations.

How much water it soaks up is only one problem surrounding cane. The other is a series of associated factors, a web of human interventions without a simple cause–effect narrative but contributing to central questions for future generations. In Chaklamba alone, farmers would not be over-extracting groundwater if they had a local water-conservation structure and better planning; if the big dam's benefits could have percolated to these farmers through some decentralized planning and execution; or if an initiative by way of providing some form of irrigation may have saved their trees, recharged their wells and kept their traditional cropping patterns intact.

Krishna Khedkar, thirty-five, one of Chaklamba's protesting farmers, has made multiple trips to Mumbai to meet ministers and legislators in the hope of one of them supporting his pleas for a sub-canal from Jayakwadi Dam to their cluster of villages. He coordinated with about seventy-five other villages in their belt for a series of identical gram panchayat resolutions with the same demand. Khedkar carried all their resolutions to Mumbai to submit to the then leader of opposition in the legislative council, Dhananjay Munde of the NCP. Munde belongs to Khedkar's district, Beed, so the farmer was particularly hopeful. But while his official paperwork mentioned the villagers' demand for their right to access the waters of Jayakwadi, Khedkar said, deep in his heart, he was more concerned about whether water is running out and whether his eight-year-old daughter's generation would even be able to practise farming.

'Farmers are actually very good environmentalists,' Khedkar says in Mumbai one rainy evening in July 2019, enjoying the respite from the still-dry Georai and Chaklamba. 'The earth is our mother; we worship our land. And yet, you're now seeing farmers drill multiple borewells deeper and deeper into her belly.' If he gets the sub-canal, would he grow cane?

He hasn't thought about it, but it seems like a no-brainer given the mounting losses on his farmland over the past five or six years. What he has spent many hours brooding over is whether the dam itself would be dry by the time the canal comes to Chaklamba. To him, selecting a crop that pretty much guarantees an adverse impact on the region's long-term water availability is a sad sign of the times, of the demoralization of a class of once-hardy people, as much a symptom of a deep malaise in Marathwada as is the spate of farmer suicides.

NINE

RESISTANCE BY THE OTHER HALF

~

'I have never seen you
Wear one of those gold-bordered saris.'

—From 'Mother', by Jyoti Lanjewar

EVERYWHERE SHE FLOWS in Marathwada, the mighty Godavari is revered as Goda Maai, the mother who gives life, nurtures and, occasionally, allows a destructive fury to slip through. Goddesses reign over temples across these eight districts. At public agitations and protest marches, a slogan will always invoke the blessings of Aai Bhavani, another mother goddess. Parbhani owes its name to a goddess, Prabhavati. And so it follows, quite naturally, that Marathwada's women are highly visible, with a collective fondness for saris in flaming pink, yellow, coral and carmine, their nose pins set with pearls and glinting stones matching the large vermilion dots on their foreheads.

Even dressed down for fieldwork, sporting scruffy shirts over saris, they are a prominent presence throughout the crop season, participating in everything from weeding the ground to tying the harvest into bundles, from tending to farm animals to fetching and storing water. At construction sites where the law-mandated hundred days of annual employment is supposed to be made available to villagers who seek work under the Mahatma Gandhi National Rural Employment Guarantee Act, women are seen wielding pickaxes and carrying headloads as much as the men. Panchayati Raj or local self-government reform has also mandated that one half of elected representatives at the village level must be women, so women sarpanchs can be seen everywhere, even if the self-anointed sarpanch-pati or husband-of-the-sarpanch is also now an uncontested part of decision-making in villages.

And yet, elsewhere, women disappear without a trace. Most prominently, they are absent from the lists of farmer suicides and landowners. These two are intertwined narratives, for women's suicides are not treated as farmer suicides because they are rarely legal owners of the land they till and care for. In June 2019, after years of activism and advocacy, and a few months after the Maharashtra State Women's Commission submitted a report on the subject, the state government issued an important government resolution for these invisible women. It pertained to ownership of farmland for widows of farmers who took their own lives.

The move came after the state women's commission held two-day-long discussions in Nagpur and Aurangabad, the respective divisional headquarters for Vidarbha and Marathwada, where officials and experts interacted with women, farm widows and activists. At great length, the women talked about the enormous weight of economic dependence after the death of the husband. Some confided about families and relatives unwilling to part with land rights or any manner of financial rights for widows, and about destitution and abandonment. They spoke of the difficulties in raising a daughter alone with no financial backing to provide for her higher education. Even continuing to till their land was a complex problem sometimes, in the face of claims over it by relatives or creditors.

The government's directives that emerged were to various departments, all with a view to ensuring the rehabilitation of and assistance to farm widows. Among the steps to be taken by the revenue and forest departments was listed: 'to transfer to the widow's name the 7/12 land extract'. This *saat-baarah*, as it is called, is a land revenue department document that shows the ownership of a plot, any liabilities on it, including if it is mortgaged against a loan, information about any disputed claim over it and a list of mutations if it changed hands from one owner to another in the past. The government resolution went on to say that farm widows who face hurdles in availing of the benefit of a house under the Pradhan Mantri Awas Yojana should be assisted immediately in obtaining the land title. Under the Rajaswa Abhiyan for quick and efficient governance, officials were to hold camps for those seeking to clarify land inheritance rights, thus also helping women get land rights.

Other government departments received instructions too. The women and child development (WCD) department would have to intervene in cases of farmer suicides to ensure that the widow receives a rightful share of property, to help such families in obtaining government aid and to train officials from across government departments to sensitize them about women's rights. The WCD department was also entrusted with the task of setting up district-level aid centres for farm widows where they could access guidance on legalities and on various government schemes. Further, the department of rural development would prioritize aid to farm widows for constructing homes under the existing Gharkul scheme. Not all instructions were perfectly clear, though. For example, the Public Health department was to 'formulate schemes' to tackle any finance crunch farm widows may face in accessing treatment for illness.

The government's move was welcomed heartily by women's groups, farmer leaders and farm widows – whoever heard of it. But four to five months later, it was impossible to estimate how effectively the resolution was being implemented. The government has published no data on the progress of this important reform, and the only way to gather this information would be to ask the *talathi*, a revenue department employee, of every village to analyse every land ownership mutation record and then to verify how many mutations were on account of a sale or a gift deed and how many due to succession or inheritance. Every case of mutation would have to be manually checked to record if it was an instance of a death, and then if the death was a suicide.

Anecdotal evidence, however, suggests that the progress is slow, if at all. District-level officials point to some issues that are not addressed by the government resolution: Are a dead farmer's sisters' rights over land to be recorded too? If a farm widow is not demanding the mutation, whether on account of domestic pressures or otherwise, should it be enforced? What if the father of a farmer who killed himself is recommending a mutation to make a minor grandson the owner instead, and what if the farm widow acquiesces to this?

These are common enough occurrences everywhere in rural India, and, doubtless, government regulations alone will not engineer the societal change that is a prerequisite for women to be considered equal partners in ownership of land and other property. Moreover, the Hindu Succession Act and a clutch of other pre-existing laws already provide property inheritance rules for widows, and these apply to farm widows too. As one Zilla Parishad CEO said, there was nothing stopping officials from promoting the implementation of land rights for farm widows, if they so wanted.

And yet, despite the laws and the latest resolution, women farmers remain largely invisible in policy and execution. The exclusion is hardly limited to Marathwada, or even to Maharashtra. The census itself calls 3.6 crore women in India cultivators, but a very large majority of these women are not considered 'farmers', who may then seek benefits such as agricultural credit, crop insurance, ex gratia for crop loss owing to rain, hail, moisture stress or pest attacks. These women find that all government schemes and subsidies for farmers are inaccessible to them. The India Human Development Survey (IDHS), conducted jointly by the University of Maryland along with the National Council of Applied Economic Research, a multi-topic survey of 41,554 households in diverse neighbourhoods of the country, found in 2018 that while women account for over 40 per cent of the agricultural labour force in the country, they own only 2 per cent of agricultural land. The IHDS conducted one round of interviews in 2004–05, and another in 2011–12. It further said that 83 per cent of agricultural land inheritance remains in favour of men.

In Osmanabad's Tuljapur taluka, Devsingha village is home to the district's first postwoman, Archana Bhosale. The matronly forty-seven-year-old is also the first woman in her village to own more land than her husband – of their seven and a half acres, she owns five. A child bride at the age of ten, when she was in Class IV, Archana's parents packed her off to her marital home in Devsingha soon after she attained puberty. She was then in Class IX. By the time she was in Class X, she was pregnant, but took her matriculation exams and passed anyway. She finally dropped out in Class XII when she got pregnant again, but later decided to complete her education and signed up for a bachelor's degree in humanities from an open university. She was in her final year in 2019, along with her youngest son, also in his final year of undergraduate studies. Her two older daughters were married, both with post-graduate degrees.

Archana had been working with a women's small savings group in the village, which was also trying to undertake a programme for women to till small plots of land independently, farm plots where women would decide what to grow and how to grow it, using only organic farming techniques. Then, around 2010, when the women farmers tried to get linked with the state's Agricultural Technology Management Agency (ATMA), a district-level society that assists farmers with technical know-how from the latest research and state-aided extension schemes for marketing or branding, they were rebuffed. They were not legal landowners. This set off a movement of sorts in Devsingha that year, with several women collectively making the demand for land rights.

The resistance was stiff. 'Even in those few households where women were treated very well, given a great deal of respect, in highly cultured and educated families, the first question to the proposition that some land be transferred to our names was – why incur the expense of registering a gift deed in favour of the woman? All of us were told we are most certainly equal owners of our family's land, but nobody was willing to put that on paper,' Archana recounts.

Archana herself had already had a brush with an inheritance struggle. Her father, an employee of the postal department in her native Salgara Devti village, died in 1993. As per rules, a family member could get a job in the department on compassionate grounds, but Archana was only sixteen and her brother even younger. Neither the family nor the department appeared keen that she take the job, but she was resolute. Starting in 1995, when she turned eighteen, she fought for years to get the job, finally succeeding in 2002, by when she had already begun working closely with women's groups.

'Women must have an identity of their own too, for it is not confidence alone that one gains from owning land. We can seek the benefit of government schemes if we are landowners; we can get agricultural loans or government compensation paid for a failed crop. Maybe when we grow old our sons or daughters will not want to care for us, but senior citizens cannot be thrown out of their homes if they own property,' Archana reasons. 'For the family too, and for society in general, it is rewarding for women to own land. Women very rarely make rash decisions about property – they're careful, they won't just sell off or mortgage a parcel for a quick fix to cash worries.'

Meanwhile, in 2011, women farmers were reaping gold in Devsingha. Organic farming meant they were improving the soil's quality, bringing home fresh produce that the family could eat as well as cereals and grain that could be stocked for most of the year. Sowing food crops as against cash crops also meant lower input costs and humble water budgets. 'We were showing results in our health, our family's health, our financial health and the soil's health,' Archana says. It took a long time, and it took hard selling of the idea that women landowners would bring home additional state aid by accessing government schemes, but finally about 70 per cent of Devsingha's women today own some land, a staggering achievement.

Against this backdrop, the June 2019 order by the Maharashtra government may be seen as a work in progress. But some conclusion about its current and anticipated impact can perhaps be drawn from previous orders on the same subject, which made similar politically correct noises on the matter of women's land rights. Like the 2019 government resolution, a 1994 one asked for state agencies to promote rural women's ownership of land and homes, but the government did not back this up with a campaign or Lok Adalat, nor did it employ any other means of locating potential beneficiaries. In 2013, the Mahila Arthik Vikas Mahamandal (MAVIM), a Maharashtra government initiative to promote rural women's empowerment, undertook a campaign based on this resolution to make women aware of and obtain their rights to a share of the husband's property.

In about two dozen villages of Parbhani district, nearly a thousand women became joint owners of farmland that year. Named the 'Ghar Doghanche Abhiyan', or 'home for both' campaign, the initiative held wide-ranging changes for rural women. MAVIM's own report on their Parbhani experiment said the women beneficiaries found they were now able to resist domestic abuse, and they had a sense of security that they could not be abandoned or thrown out of their homes. Still, for all its success, the Ghar Doghanche Abhiyan was never implemented in other districts, nor was it scaled up. The state government gave no explanation for quietly forgetting a progressive, pro-women campaign that had raised men's hackles across the region where it was implemented.

So nobody would have been unduly surprised when about five years later, in November 2018, seven months before the government's latest order on land rights for farm widows, a survey by the Mahila Kisan Adhikar Manch, a federation of women's rights groups from Vidarbha and Marathwada regions, found that of 505 farm widows surveyed, about a third had been unable to get their farmland legally transferred to their name. Forty-one per cent of the 505 women did not have an independent ration card and remained dependent on their marital families for a share of the highly subsidized public distribution system (PDS) foodgrain. A third did not know of the existence of a pension scheme for widows. Among those surveyed, sixty-nine families had needed money in the past couple of years for medical expenses, and forty-seven families had taken a loan for the purpose. Only 19 per cent of those surveyed knew of the state government's flagship health insurance scheme.

~

CENTRAL AS IT is to a woman's life in rural India, land is still only one part. As cyclical drought and its attendant problems of deficiencies and deprivation tear through the Marathwada countryside, the personal agonies of one half of its people remain shrouded within homes.

On the morning of Dussehra in 2014, Anita** realized in the very instant she saw her husband hanging from a roof beam inside their home in Osmanabad's Tuljapur taluka that her life would never be the same again. Her main concern was her two boys, then aged five and two, and a source of income that would be steadier than the plot of *gairan* land she and her husband had been cultivating until then. A decades-long crusade for the Dalit occupiers of such land to obtain legal titles remains incomplete, so women such as Anita are not entitled to the government's ex gratia for farm widows, because technically her husband was a landless labourer. According to the state's narrow definition, his suicide was not on account of farm distress. In addition to her sudden and total dependence on her father-in-law, there were creditors to face, and a male relative's repeated sexual advances in the months that followed.

** Second name withheld to protect her identity.

Even among the somewhat better off, the years of drought and overwhelming household anxiety about finances precipitated another silent crisis, of women's health and nutrition. Within that spectrum, reproductive health elicits the maximum neglect. Just how, women ask, can they consider spending a few hundred rupees on a doctor because they are bleeding for two weeks or because their monthly period now comes with a debilitating backache. So, in the summer of 2016 when Pankaja Munde of the BJP, the then MLA from Parli in Beed district, announced a free medical camp in association with some volunteer doctors and the Beed Civil Hospital, hundreds of women signed up immediately for long-postponed medical procedures.

A huge majority could be found in the surgery and gynaecology wards. Among them was Kesarbai Ghodke and her daughter-in-law, Sonali Ghodke, from Ghodkarajuri village, six kilometres away from Beed city. Kesarbai had ignored the shooting pain in her abdomen for nine months straight, and said she preferred to try home remedies so that she could save the medical expenses. Their farmland had suffered losses since the crippling drought of 2014, and two years later their savings were almost entirely gone. Sonali and her husband used to run a computer class in Beed city, but business had dried up along with the failed monsoon.

The Ghodkes' story was strikingly similar to tales in every ward of the dank, fusty government hospital. Weddings had been called off suddenly or postponed, girls were pulling out of college and even high school, the older boys were away in the bigger cities looking for work, a few of them were living in shanty towns in Pune and Mumbai, and doctors said anaemia and vitamin deficiencies were common on almost every medical chart. As Marathwada's economy took another big blow in its third drought between 2012 and 2016, the growing sense of doom silently consumed women's health and girls' higher education. In fact, the state has not even begun to study the social impact of Marathwada's cyclical drought.

On 1 June 2015, Kalyan Khomne, fifty-five at the time, hanged himself from a neem tree on his little plot of land in Khomnewadi, just outside Nandurghat village in Beed. Their home, two rooms of unplastered brick with a cemented verandah, abuts the two-acre land. Kalyan was discovered quickly, but it was still too late. In the preceding months, Nandurghat village and nearby hamlets around Khomnewadi had witnessed five farmer suicides. One day, a week before he became the sixth, Kalyan read out a newspaper report to his son Shahdev. It was about a farmer's suicide in their taluka, Kaij, one of the worst-affected regions during every drought. Kalyan told Shahdev the family had received ₹1 lakh in compensation from the state government. Kalyan had loans totalling nearly ₹3 lakh.

More than four years later, Kalyan's wife, Chandrabhaga, submitted applications for the state-sponsored widow's pension and tried to ask officials if she might be eligible for any other state scheme, but nothing worked. In 2014, the state government had set up the Vasantrao Naik Shetkari Swavalamban Mission, a project to make farmers independent. In 2016, it set up the Prerana Prakalp project to tackle mental health issues in districts where the incidence of farmer suicides was high. Private moneylenders were being regulated under a new law. A bevy of new financial aid schemes appeared, for widows and children of farmers who take their lives. But as far as Chandrabhaga could see, multiple initiatives by the state, well-intentioned and elaborate as they seemed on paper, fell short of providing her relief during the blistering summers of loss.

Chandrabhaga, named after a river to celebrate its abundant gifts, knew of the Sanjay Gandhi Niradhar Yojana for abandoned or widowed women. It is only ₹600 per month, but she would be grateful to have it. She did not know about the Mahatma Phule Jan Arogya Abhiyan, the state's very successful health insurance scheme. She had also never heard of the Prerana Prakalp project. As for the ex gratia that had played on her husband's mind in the last few days before he died, they got the money as well as some aid from an NGO, and almost all of it was spent within months, repaying loans. The leftover was spent sowing a rabi crop that year, only to see it fail.

Still, the four intervening years had returned Chandrabhaga's smile to her wan face. Shahdev was more mature, and the family had found once again its old rhythms of hunting for work and credit and making small gains on their land before another round of losses. In 2015, all Chandrabhaga could bring herself to tell journalists was that her husband spoke little; he had always been a bit gruff. He had never discussed his inner demons with his wife. Being alone for years toughened her up, and in the summer of 2019 she said confidently, 'I've submitted my documents asking for help; maybe the government does not have any scheme for me.' Chandrabhaga made the thirty-kilometre trip to the taluka office twice and returned home empty-handed both times. Eventually, Nandurghat journalist Amol Jadhav intervened, and when a public sector company mentioned it was donating sewing machines to farm widows, he made sure Chandrabhaga was among the recipients.

For an unlettered farm widow, a visit to a government office, carrying papers she cannot read, signing or placing her thumbprint on further documents, trying to speak coherently to a government officer, or just even negotiating the journey into that office would have been inconceivable until the tragedy. Pushed into the deep end by circumstances, women like Chandrabhaga have had to overcome their own fears and the very real threat of violence or sexual assault during an unfamiliar journey, and they often return home in greater despair like she did. Activists have all suggested a monthly camp in every taluka to sort out common problems that farm widows face, and a similar 'mission mode' approach on land rights for farm widows. The simpler, effective solutions are just never dramatic enough to be announced with fanfare by governments, they complain. To compound her tragedy, Chandrabhaga's other son, Sambhaji, took his life in June 2020, frustrated with the family's continuing troubles. She was back, emotionally and financially, where she had been five years ago.

~

'ONE EVENING, A river named Godavari came rushing into my home.' That is how Shailaja Narwade, a strapping mother of three in her mid-forties and soon-to-be-president of a rare all-women Farmer Producer Company in Osmanabad district's Mhasla village, introduces her mentor, Godavari Kshirsagar. Godavari has spent nearly two decades organizing thousands of village women in Latur and Osmanabad districts into small savings groups, simultaneously providing them training in organic farming practices and enabling them to get credit to set up small businesses.

Early in 2000, Godavari, having lost her husband in a road accident a couple of years ago, was battling shock and depression at her mother's home in Gandora village in Osmanabad's Tuljapur taluka. She had spent nearly a year indoors, cycling between listlessness and panic at having to bring up her boys, then aged four and one, alone. Nagged by a concerned family to step outdoors and socialize, she joined a women's self-help group and began attending its meetings. She was among the few women in the group who had had some manner of formal education, having attended school till Class VII. So she slowly began to handle the group's paperwork and accounts, eventually going on to establish more women's small savings groups in Gandora.

The '*bachat-gat*', that is, the small savings or self-help group, is a legacy of the rehabilitation works in and around Latur, where 8,000 people died in an earthquake in 1993. Gandora, incidentally, was one of the new villages built after 1993. Swayam Shikshan Prayog (SSP), an NGO, was established in 1998, but its early origins are in the community-led rehabilitation effort following the Latur quake. 'The Latur crisis and reconstruction of houses gave locals the opportunity and impetus to mobilize rural women for large, community-centred efforts. In time, SSP mobilized these women into self-help groups,' says Prema Gopalan, executive director of SSP. The focus of the relocation and rehabilitation was not on safe buildings alone but also on community participation. The community-based organizations set up alongside the relocation helped women organize for economic empowerment and legal awareness, the self-help and savings groups proving to be a means for women to make some savings and access credit.

Godavari is a key resource person at SSP. Less than two decades since she first stepped out of Gandora to coax women in other villages to form self-help groups and to train them, she has travelled to half a dozen countries to share her incredible story of what women could achieve in Latur and Osmanabad districts' villages when they mobilized and took on a more assertive role at home and on their farms. 'In every drought-hit village in the region where we work, women now play a significant role in running the household. They shoulder major responsibilities because they're able to replace a lost income, whether it is through a home business or through organic farming on the one-acre model that we promote. And the natural next steps are to get land rights, insist on higher education for girls and make sure that all women have adequate opportunity to earn an income, so this change is growing quite organically,' Godavari says.

This one-acre model is where Shailaja Narvade began her journey with women's activism too. In 2003, in the decade that she had lived in Mhasla village since her wedding, the Narvades had seen multiple droughts, but that year's losses and scarcity were boundless, she says. The family's agricultural income dried up entirely. 'At one point, there was nothing to eat at home. Neighbours gave us food,' she recollects. As a land-owning caste, they had never worked on others' farms or at employment guarantee scheme sites, but between wounded pride and living on charity, it was an easy enough choice. 'It was my mother-in-law who took my hand one day and said we couldn't go on living off charity. The next morning, we – my mother-in-law and I – got dressed in our best saris and went to work breaking stones at a site where they were building a water storage structure.'

The wages the women received were under Maharashtra's employment guarantee scheme, the precursor to the all-India law guaranteeing a hundred days of work on demand each year. Back home, she would look at her blistered palms and sob, but the duo worked as labourers right through the latter half of 2003. In 2004, the rains were kinder, and she met Godavari and the other women from SSP. She led the formation of a women's savings group in Mhasla – twenty women at first, then forty. As the group grew, the women underwent training in organic farming and launched the Sakhi Anna Suraksha initiative, a food security programme, in the village, under which women farmers take charge of a one-acre plot of the family's landholding and grow a combination of food crops that the family can use at home through the year. This includes sorghum, millet, various vegetables, lentils, turmeric and leafy greens. All of this is grown organically without any chemical fertilizer, instead using compost pits that the women collectively or individually plan and maintain.

A couple of years later, Shailaja borrowed money from the savings group to buy milch animals, adding that income to her home. By 2012, with the family well on its way out of their 2003 crisis, she was in a position to request, and get, two and a half acres of their family land legally transferred to her name. Her son is now twenty-one and is planning his own entrepreneurial initiative in Mhasla, while both daughters are pursuing higher education outside the village. All three children know how to run a farm and are perfectly comfortable with the hard labour it entails.

The one-acre model, now benefiting over one lakh women and their families, is promoted by SSP as a climate-resilient farm practice. It urges women to seek cultivation rights over a half- or one-acre patch of the family's landholding to grow multiple food crops for home consumption, while selecting the most water-efficient and short-term crops, emphasizing water conservation through the use of hydroponics, drip irrigation, sprinklers, farm ponds, recharging of borewells, tree plantation, etc. It also builds resilience among small and marginal farmers using biofertilizers and the preservation and exchange of local seeds. It views women as bearers of knowledge, enabling them to make informed decisions on what to grow, what to consume, how much to sell and where. 'It promotes food, nutrition, income and water security,' says Gopalan. And because the model encourages women to obtain cultivation rights, in all these villages the women have initiated in their homes the sticky conversation about land rights.

Gopalan says SSP began to work on promoting organic farming and the Sakhi Anna Suraksha programme several years ago, but around 2014 they evolved the approach with the specific target of building leadership among women who could then work on ensuring food and income security in their households and among other women's households too. Significantly, the second part of the programme is to promote agrobusinesses among Marathwada's women and add an income to the household as a shock absorber for climate-related farm income fluctuations.

Ghatangri village is located off the smooth and scenic Osmanabad–Barshi highway, past a shimmering lake and the Dharashiv caves. The rock-cut caves are a former Buddhist and Jain site of worship dating back to the era of the Rashtrakutas, who were among the rulers of the Deccan before the Bahmani Sultans and the Nizam. Archana Thorat and her husband own only half an acre of land in Ghatangri, and it is stony, unrelenting and entirely uncultivable. Archana sought a loan from a fund operated by SSP for small businesses and started her own poultry farm alongside a couple of milch animals. Her biggest success, she says, is putting behind her, for good, her years as an agricultural labourer. Her husband still works on other people's farms while she runs the poultry business. Also in Ghatangri, a group of fifteen women, all with extremely small landholdings, got together to start a floriculture business with a loan from SSP. Their marigolds and gaillardias now make their way to the markets in Osmanabad town at the crack of dawn every day, their joint venture helping build scalability and cutting costs through a drip irrigation system. These loans, critical to tackle early risk, come from what is now a ₹1 crore pool built by SSP with seed money from a corporate, from which loans of ₹10,000 to ₹30,000 are given at a 12 per cent rate of interest.

Poultry farming was a popular business among the women who took these loans, many especially keen to rear the Kadaknath, an expensive, all-black chicken native to central India and considered a superbird for the distinctive taste and health benefits of its dark meat. In Kalamb taluka's Janpur village, Sheetal Ambekar, who took a loan to set up a unit for manufacturing *khowa*, says the ramifications of village women getting these agrobusiness loans were wider than they anticipated. 'For one, families just stopped taking money from the sahukars (private moneylenders),' she grins. Across the villages, where self-help groups run under the umbrella of SSP or Umed, the state government initiative of the Maharashtra State Rural Livelihoods Mission, the women say that at least 50 per cent of private moneylenders have gone out of business.

Further, in Janpur, a modest village of about 300 families, there are at least twenty-five women-run enterprises today, including grocery-cum-cosmetics stores, flour mills, papad-making units and poultry and milk produce businesses. Sheetal and her husband used to together earn approximately ₹300–350 per day as daily wagers, when work was available. For two years now, neither has had to look for work as labourers, their monthly income a steady ₹25,000–30,000. One of their two boys has moved from a school where the medium of instruction was Marathi to a semi-English-medium school. Janpur is one of about twenty small hamlets clustered around Pangri in Kalamb, Osmanabad, and women from these hamlets have similar success stories. One woman, Anusaya Gore of Naigaon village, says she is expanding her incense-stick-making enterprise and has sought a ₹1 lakh government-backed loan offered to non-farm enterprises under the Prime Minister Mudra Yojana.

These women laugh easily, at least a couple of them wear eyeliner with fashionable winged tips, the younger women keep their earphones plugged in to take calls or to listen to Bollywood songs. When they are required to attend taluka-level group meetings, virtually everyone selects saris of polyester chiffon that do not crumple easily during the bumpy ride in a state transport bus. Some ride two-wheelers and have scarves handy to fend off the dust on the roads. All of them truly defy the stereotype of poor, helpless women. Yogita Jagtap of Ashu village in Ausa, Latur, has a master's degree in history and a BEd, but found herself suddenly homebound after marriage until she started a packaged foods enterprise through a loan from SSP.

Simultaneously, these women have all been small and marginal farmers or landless labourers. They have witnessed multiple tragedies on their farmland, depleting groundwater, wells running dry, cycles of bad loans and the all-consuming stress at home. 'Many of us have found increasingly that our husbands are depressed, feeling no hope. Or they're just angry, irritable all the time. And simultaneously, more and more women are running families; everybody acknowledges it now,' says Archana Koli, a mother of two in Salgara Devti village, Tuljapur, who has a successful poultry farm.

Chetna Sinha-Gala, whose Mann Deshi Foundation operates the state's first cooperative bank for women that was started as a response to rural women's lack of access to credit, says there are clear linkages between putting money into a woman's hand and long-term family prosperity, a model that the Mann Deshi Bank has helped create in Western Maharashtra's Satara, where it operates. But she says access to credit for rural women is an area that needs sustained research by banking institutions. 'Mann Deshi offers a model, but there is still so much to understand about a range of possibilities between microfinance and the Mudra loans for micro-units, for example.'

Women, as opposed to men, are an ideal target for forming self-help groups and small savings groups, she says, because the model works on peer pressure to collect the savings and make loan repayments, which is almost never possible to build in a group of men. Also, according to Sinha-Gala, men are not enthusiastic about small loans. 'They want to play with big, mainstream sums,' and, in addition, women shoulder a different kind of responsibility with regard to their families, their children's well-being, educational expenses, medical bills, etc. This leads them to be more regular and reliable members in a small savings group.

But one spin-off from this otherwise happy story is that, in diverse circumstances, the results of giving women loans may vary sharply. The long drought has led women in many pockets of Marathwada to begin to shoulder greater financial responsibilities, but this also means women in the region are now in debt, sometimes unsustainably so, in a clear departure from the previous decade. In village after village in the slightly better-developed districts of Aurangabad and Jalna, it is apparent now that most women have at least one loan against their names, sometimes more.

When the Sri Krantisinh Bahuudeshiya Sevabhavi Sanstha, a multipurpose charitable organization, conducted a survey in Jalna's Ramnagar village, they found that a large majority of women now had at least one loan to their names and most said this had not been the case a few years earlier. For non-banking finance companies and micro-lending companies, the women of Marathwada were a virgin marketplace when they arrived about a decade ago. Especially in villages closer to towns where these companies have offices, thousands of women now have loans ranging up to ₹50,000, taken as personal loans and used to tide over a household finance crunch, for a daughter's wedding or college fees – not always used as capital for productive business, nor as credit support for sanitation infrastructure, a major push area. Usha Shinde of the Sanstha, which runs a skills training centre for young girls in Ramnagar as well as self-help groups for women, says up to 70 per cent of the women they surveyed had taken loans up to ₹1 lakh. 'We surveyed 682 families – almost all of Ramnagar's population of 3,250. We spoke to 1,500 women, and approximately 1,100 had taken loans from small finance companies,' she says.

Suman Raut of the Prerana Self-Help Group in Ramnagar says she took ₹80,000 from a microfinance company for her daughter's education. The girl is pursuing an MCom degree in Ahmednagar city. In addition, Suman took ₹5,000 from the small savings group. Suman would have liked to take the loan for an agrobusiness instead, perhaps a set of milch cows. 'But we needed the money then; we're repaying slowly.' She has no source of income of her own.

In Ganeshpur, part of the Manegaon group panchayat in Jalna, fifty-something Arjun Pitale jumped into the well outside his home in February 2019. Ganeshpur, a remote hamlet with just 135 families, had suffered a dreadful water scarcity through that year's monsoon, and his one-acre plot was no different. The family said he managed to harvest one and a half quintals of cotton in total during the previous harvest; average cotton yield in their village in a decent year was about ten to eleven quintals from an acre. Pitale usually made ends meet by also working in the home and fields of a large landowner in the village, and late in 2018 he took ₹70,000 from this man as advance payment for a year's labour. That money was spent on his elder daughter's wedding. Pitale's father had an outstanding bank loan too, about ₹50,000, which the son was expected to help repay. Together, the farm losses and the loans piling up had made him moody and despondent by 2019. It was a familiar story. In addition, Pitale's wife had borrowed ₹60,000. It was ostensibly a loan for a home business, but the money was eventually directed towards the wedding expenses. With him gone – leaving behind their younger daughter, seventeen and still to be married, and a younger son with special needs – his wife was inconsolable.

About the glut of loans for women in Marathwada, Sinha-Gala agrees that there may be an overborrowing problem. Among the all-women customers at Mann Deshi Bank applying for loans, a look at their CIBIL credit scores shows 30 per cent have overborrowed and are lagging in repayments. 'Credit on its own is not the devil,' says Sinha-Gala, who was one of six global women leaders selected as co-chairs at the World Economic Forum, Davos, in 2018. 'The problem is that the decision-making power of how much credit is to be taken is not sufficiently in the hands of the woman.'

The aggressive marketing by microcredit institutions in rural Maharashtra means the loan market for a woman in a Marathwada village is entirely dictated by the supply side, including the loan amount, the period of repayment and conditions. The supply side is so aggressive that third and fourth loans, usually to repay the first loan, are common. 'This is bad because a woman is now paying double the interest,' says Sinha-Gala. 'In other sectors too, market forces are decisive but in circumstances where resources are so scarce, as in a drought-hit region, this becomes worrisome.'

On the other side, while microcredit is growing rapidly, micro-savings is nowhere near as aggressive. Mann Deshi, for example, offers doorstep collection of small savings, the bank bearing the cost for this. Because these are typically very small sums, providing services for last-mile savings is a cost that banks simply avoid. Sinha-Gala says, 'In the villages,

I see women with two or three loans each, but never a woman with two or three savings accounts. This is despite the fact that as a society we are a people who save. In poor families too, women will put away tiny sums they have saved into grocery jars hidden away somewhere.'

Even the Ujjwala account under a central government scheme, a direct benefit transfer account for a government subsidy, is jointly held by a woman and her husband; most women will never put their savings into an account where they do not solely control withdrawals. The third aspect of women's growing indebtedness is the absence of any form of debt counselling. Whether it is the wife of the Ganeshpur farmer who jumped into a well or the Ramnagar savings group member, most women who have taken loans have inadequate financial literacy.

Despite the glitches and the slow pace of change, however, the women of Osmanabad and Latur who run their own small-scale businesses today want to assert that they are no pushovers, not even lucky exceptions. They represent the best success stories of a troubled land, their model based on backbreaking work with the odds stacked against them. They speak with grace and erudition about what they have won – better bargaining power at home and equal decision-making rights.

They have also smoothly made space for themselves at the big boys' table – Shailaja is president of the Vijayalakshmi Farmer Producer Company (FPC) in Mhasla, Tuljapur. In 2016–17, the government administration in Osmanabad formed thousands of self-help groups and more than sixty FPCs, of which Vijayalakshmi is the sole all-women company, readying to stock and sell organic *urad* dal produced by fifty farmers, most of them women. Shailaja now works with 4,000 women in four villages around Mhasla, as a taluka-level leader–representative of the women's small savings groups.

'Every woman should till two acres of land; that's my recommendation now,' says Shailaja, who has given speeches at village events on women's equal participation. She says she learnt what leadership really means through her years as a farmer and activist. 'I had studied till Class XI when I got married in 1993, but the real learning I have had is everything I gained since 2004. What I have that the men do not is the quality of leadership without arrogance. And in that is a solution to many of our problems.'

T E N

THE IMPOSSIBLE
DROUGHT-PROOFING DREAM

~

'If there's no water in the dam, should we urinate in it?'

—Ajit Pawar of the NCP, then deputy chief minister of
Maharashtra, in April 2013

IN 2007, AS Australians emerged from the worst phase of the Millennium Drought that ravaged parts of the country from 2000 to 2009, exposing acid sulphate soils and forcing wetland systems to be temporarily disconnected to save water, Prime Minister John Howard proposed to the country's parliament what would be their biggest-ever water reform, one that would cost billions of dollars in taxpayer money and a project that would quickly become the centre of his re-election campaign. Australia was enduring its worst-ever drought in history, inflows into the Murray River had hit a record low, and the prime minister wanted the federal government to take over the management and operation of the Murray–Darling Basin

174

water systems. The Water Act followed, and then the Water Regulations the next year.

The Act itself would be revised in later years, with periodical reviews planned in order to continually improve laws for sustainable use of resources. In recent years, the evolution of those reforms has faced sharp criticism, but in 2007 it was a pathbreaking project. For the very first time in a country where some regions had witnessed a thirteen-year drought, the law established that water planning for the entire Murray–Darling river basin would be undertaken as a whole, instead of deciding state by state. Empowering the Murray–Darling Basin Authority to do so was key to strategic, integrated and sustainable management of water available to Australians, the law held. 'This is our great opportunity to fix a great national problem,' Howard told the nation. 'It can only be solved if we surmount our parochial differences.'

Disparate as federal administrative systems in Australia and India are, surmounting parochial interests will be key for Marathwada too, for an equally contested and expensive plan, audacious by its sheer scale, is proposed as a central long-term solution to the region's chronic water shortage. The plan is to transfer water from the surplus river basins of the Konkan to the deficient basin of Marathwada. The Konkan is the sea-kissing strip of districts bounded by the Sahyadri mountains on its landward side, bringing it plentiful monsoons, seasonal and perennial rivers. While some talukas in Marathwada receive an average seasonal rainfall of only 400 millimetres, spots in the Western Ghats or the Sahyadri range just a few hundred kilometres away could receive up to 6,000 millimetres of rain in the same four-month season.

In Maharashtra, the number of rainy days during the monsoon ranges from an average of forty in parts of Marathwada to up to a hundred in the Konkan. As a result, 45 per cent of Maharashtra's water resources are locked in the monsoon rivers that tear down from the Sahyadris. But the average height of towns and villages west of the ghats is sixty metres above sea level, and the average height of these basalt mountaintops is 600 metres. Whatever water cannot be harnessed by human engineering as it comes crashing down from those heights just drains into the Arabian Sea.

The Godavari River is traditionally considered a normal basin, while the Tapi basin to its north is deficient. A normal basin typically provides 3,000 cubic metres of water per hectare of land every year via reservoirs and canals, in addition to rains received by farmland. Marathwada, with its repeated cycles of very poor precipitation and depleting groundwater levels, is no longer seen as adequately provided for by water available in the Godavari basin. Of the five river basins in Maharashtra, one is formed by multiple west-flowing rivers. The other four, the Tapi, Narmada, Godavari and Krishna basins, account for 92 per cent of Maharashtra's cultivable land and more than 60 per cent of its rural population. Nearly half of the area of these basins is now considered deficit or highly deficit, and deteriorating as demands on the scarce resource grow rapidly.

On 27 July 2019, when nearly 1,500 people were stranded on the Mumbai–Kolhapur Mahalaxmi Express that was marooned in floodwaters less than sixty kilometres from Mumbai, on board the train was a rather thoughtful Rajnish Ramkishor Shukla, deputy secretary to the Government of Maharashtra. The townships of Kalyan, Ulhasnagar, Murbad and Ambarnath, all located in the outbacks of Mumbai, had witnessed torrential rains since the previous night. Murbad had received an average of 332 millimetres of rainfall in twenty-four hours, Ulhasnagar had received 296 millimetres. These towns stand along the Kalu,

Ulhas, Waldhuni and Bhatsa rivers. The rapid, unplanned growth of these satellite towns of the financial capital, with a rash of encroachments and rampant illegal sand dredging, has ravaged these rivers' flood plains.

As naval choppers photographed the stranded Mahalaxmi Express, the train appeared to be floating in the Ulhas river's flood plains. As choppers and boats were pressed into rescue operations, Shukla could hardly believe what was happening. A senior official at the state's water resources department, he was scheduled to make a presentation to then Chief Minister Devendra Fadnavis in the next two days on how much excess water flows through the Ulhas River, and how much of it could be conveyed to the Godavari basin instead.

On 30 July 2019, the Maharashtra government issued a key resolution. Technical studies and detailed project reports would be prepared, it said, for various river-linking projects that would together lift 115 TMC (thousand million cubic feet) of water from the west-flowing Nar-Par, Damanganga, Vaitarna and Ulhas river basins to the parched Marathwada region.

Exciting as they appear, inter-basin water transfer projects are not exactly new. As far back as the 1970s, the National Water Grid and Garland Canal were considered and discarded as techno-economically unsuitable. Later, in the 1980s and 1990s, the National Water Development Agency (NWDA) identified thirty potential river-linking projects, including two in Maharashtra and Gujarat, with the Konkan's rivers being their common point. The two states even signed a memorandum of understanding (MoU) in 2010 for two priority links, the Damanganga–Pinjal and Par–Tapi–Narmada. Detailed project reports were prepared, but with these running into long delays, the Maharashtra government resolution of 30 July, a decision that came after Shukla's presentation on the likely success and urgency of river-linking schemes, said projects key to Marathwada's water crisis would be taken up as state projects to simplify and expedite matters.

The basins of west-flowing Ulhas, Vaitarna, Nar-Par and Damanganga rivers have an excess 360 TMC of water. Projects are to be finalized for lifting and conveying 115 TMC to the Godavari basin – the new projects will totally convey 112.23 TMC, as small water-conveying systems functioning on gravity already exist, totalling 2.73 TMC. Marathwada will receive 25.6 TMC for the time being, widely considered to be a critical addition to the Godavari basin's current water availability. Detailed project plans are ready for the Damanganga–Pinjal and the Par–Godavari links. Before 2021, the water resources department is also scheduled to have detailed plans for the Nar–Par–Girna and the Par–Ekdare–Godavari links.

But lifting water across the rock face of the Sahyadris will be complex and challenging, to say the least. Any dam construction would have to be at a sufficiently low point on one side of the ridge to benefit from optimum catchment. From here, it will have to be pumped up to a series of intermediate ponds or reservoirs before tunnels through the rock can convey it eastward. Benefit–cost ratios, pumping power requirements, plans to use solar power and fine-tuning of other technical assessments are still being carried out. State government officials have begun to call the plan a *vikas-ganga*, or a flood of development, coming to the water-scarce Marathwada.

The other major project, central to resolving Marathwada's water woes, is for drinking water, to counter the few hundred crores of rupees spent annually on supply to villagers through water tankers. This is the Marathwada Water Grid project, proposing to connect the eleven major dams located in the eight districts of Marathwada through pipelines that will range from 1.6 metres to 2.4 metres in diameter. The concept is, once again, to take water from the water-rich reservoirs to any region serviced by reservoirs that are deficient at the time.

The design is to build two loops of connecting pipelines. The first is a primary loop that will connect the plentiful reservoirs to areas serviced by reservoirs with low storage levels. For example, Nath Sagar, the reservoir alongside the region's largest dam, Jayakwadi, began to overflow in August 2019 even though several talukas in Marathwada were yet to see substantial rains. Nashik, towards the north of the state, had received torrential rains through July and August, and the famed riverside temples of Nashik were submerged by the mighty Godavari in spate. As excess water from upstream Nashik drained into Jayakwadi Dam, the 2,171 million cubic metres capacity dam filled up.

The Jayakwadi, curiously named after a village in Beed district where the original dam site was proposed, is a ten-kilometre-long structure, with a reservoir that is fifty-five kilometres long and twenty-seven kilometres wide. Its name, Nath Sagar, is apt when the reservoir is overflowing, its restless waters resembling the ocean. In August 2019, villages in Beed and Aurangabad located along the canal length of Jayakwadi Dam rejoiced as water was released from sluice gates via both left and right bank canals. At the same time, at least six of the other ten major dams of Marathwada continued to remain at dead storage.

According to the plans for the Marathwada Water Grid project, giant pipelines running thousands of kilometres will convey water from a water-rich reservoir, such as Nath Sagar, to pumping stations and water-treatment plants and from there to talukas that are water-scarce. The plan is to make optimal use of water whenever it is available in individual reservoirs but while the region remains collectively water-scarce.

The water grid will function much like a power grid according to the Maharashtra Jeevan Pradhikaran, the government agency that will execute the project through private concessionaires. Some sections of the pipeline will allow reverse flows, to optimize the system so that a water-scarce taluka is supplied from the nearest water-surplus reservoir. A secondary network of pipelines will convey water to each of Marathwada's seventy-six talukas, with a tapping point at every five to ten kilometres of pipeline. P. Velrasu, Indian Administrative Service officer and former member-secretary of the Maharashtra Jeevan Pradhikaran, believes the Marathwada Water Grid can be a potential 'game-changer' for the region. The government agency has already floated tenders for what will eventually be a ₹16,000-crore project.

The concept of the Marathwada Water Grid comes from a water supply network built in Israel, which is projected as a problem-solver for the global water crisis with its innovations in water supply. In the book *Let There Be Water: Israel's Solution for a Water-Starved World*, American entrepreneur and water activist Seth M. Siegel presents an account of how water-poor Israel applied home-grown technologies and state-of-the-art policymaking in water to go from a desert country to a 'water superpower'. Sixty per cent of Israel is desert, the rest at least partly arid, making its ability to export water to small neighbours a matter of interest for Marathwada's policymakers.

Siegel's book goes back to the 1930s, when the Zionists drew up a water plan, and the 1940s, when deep drilling located water in the Negev desert. In 1955, the Yarkon–Negev pipeline was inaugurated, part of the 130-kilometre National Water Carrier. A submerged pipeline under the northern part of the Sea of Galilee conveys water to reservoirs and pumping stations. That was one engineering feat, followed by more innovations in monitoring use and pricing of water. By the time the Government of Israel declared that it was free of weather-related water shocks, much else had been done, including pricing water at real costs, and control and management of water resources entirely by technocrats delinked from any political interests. The inspired Marathwada Water Grid uses Israel's national water company Mekorot as its technical consultant, but the troubled region's socio-political realities are far removed from much of Israel and its ingenious solutions.

Then Chief Minister Devendra Fadnavis had said Marathwada's drought would soon be 'history' once the water-linking system and the grid projects are implemented. Speaking in Beed during his Mahajanadesh Yatra, or People's Mandate Tour, a massive outreach programme launched months ahead of the 2019 assembly elections, Fadnavis said the lakhs of water storage structures built under his Jalyukt Shivar programme were a success, but continuing rain failure had rendered them inadequate. 'Under the Marathwada Water Grid, dams will be interlinked and water will be taken to villages. It will cost ₹20,000 crore,' the chief minister had said. 'Around 167 TMC water from Konkan will be lifted and transferred into the Godavari basin. These two schemes will make Marathwada's drought history.'

In October 2019, the water grid featured prominently in the election campaigns of the region's BJP candidates. Then Water Supply and Sanitation Minister Babanrao Lonikar, a legislator from Partur in Jalna, said his trump card would be the Partur–Mantha Water Grid on which work was already underway, and which would eventually become part of the larger Marathwada Water Grid project. He wasn't bluffing. The state government had initiated a plan to use water from the Lower Dudhana Dam and from wells to be sunk in its submergence area, to supply water to 173 villages in the talukas of Mantha, Partur and Jalna through this water grid. The grid would comprise 700-millimetre pipelines, pumping stations and a water-purification centre. Without the unwieldy size of the bigger project and with nimble local tendering, Lonikar crafted the water grid project into his own localized election vehicle.

The Marathwada Water Grid is not the first of such schemes. Telangana built a similar network, while Gujarat has a drinking water project that is a hybrid of pipelines and canals from Sardar Sarovar Dam. But the Maharashtra government's plan to provide piped drinking water to every village household by 2021–22 will be the most ambitious. At a later stage, the project envisages connecting to the grid waters from Konkan's west-flowing rivers that now empty freshwater into the Arabian Sea and also from the Krishna basin that keeps Western Maharashtra well-irrigated and lush. These will dovetail with the river-linking designs already drawn up.

The Maharashtra Jeevan Pradhikaran foresees little trouble in completing the project. For one, no contentious land acquisitions or negotiations with landowners are required, except for the water treatment plants. The pipelines will be laid along existing highways where the state already has right of way. Of course, some challenges remain: the ₹16,000-crore project involving private players will have to be matched by an almost equal sum from the state budget, from a government that is reeling under a debt of ₹6.7 lakh crore. But without state contribution, construction of the much more dense and complicated tertiary networks of pipelines towards villages cannot be undertaken. Also, under the hybrid annuity model, following prequalification and vetting of designs, the bidder quoting the lowest net present value (NPV) for construction and operations for fifteen years will be selected. The concessionaire raises 60 per cent of the project cost via debt and equity, while the government pays the remaining 40 per cent in tranches. This too will require budgeted revenue support from the government for the concessionaire, over fifteen years.

Meanwhile, from big-dam opponents to environmentalists and engineers, the cautionary word to both mega multi-crore projects – the river-linking system and the water grid – is that these will merely kick the water scarcity can down the road. The projects are rooted in a common philosophy – transferring water from water-rich regions to water-deficit regions. Demand in the water-rich regions will rise inevitably, and a demand–supply deficit in the water-scarce regions may grow as well. Ultimately, both seek to provide water to tackle symptomatic thirst. So, it follows that they can prove to be pivotal for Marathwada only if complemented by initiatives for sustainability. And for that, more existential questions must be faced: how do we reduce water demand, how do we recharge groundwater, how do we ensure ancient aquifers survive another 10,000 years?

~

DURING THE SOUTH-WEST monsoon of 2015–16, average rainfall in Kadwanchi village in Jalna district was only 412 millimetres. Ensconced between towns receiving drinking water supply once every fifteen to twenty days during the summer, in a region where crops suited to dryland farming are cultivated, Kadwanchi has a rare success story of water conservation and watershed management. Sarpanch Chandrakant Kshirsagar, a cheery man with a booming voice that takes on a rich timbre when he is excited, says the village's farmers collectively earned ₹27 crore in 2012–13. In 2015–16, despite being a drought year, output grew to ₹42 crore, and per capita income rose to ₹1.27 lakh. 'In the 1990s, per farmer income was barely ₹3,500 a year,' Kshirsagar says. Incredibly, Kadwanchi produces grapes, which is very unusual for arid regions because an acre of grapes requires 10 lakh litres of water during one crop season. But Kadwanchi's graceful vineyards occupy nearly 2,000 acres of land. And one of the village's biggest achievements is matching the quality of grapes produced in the vineyards of well-irrigated Nashik and parts of Western Maharashtra.

In 1995–96, the Krishi Vigyan Kendra, established by water and agriculture expert Vijay Anna Borade's Marathwada Sheti Sahaya Mandal and the Indian Council of Agricultural Research, New Delhi, decided to begin a watershed development programme in Kadwanchi, located eighteen kilometres from Jalna city. The work cost the agencies ₹1.2 crore at the time. Considering the village's geospatial factors, as well as those of neighbouring areas, the team drew up a plan to harvest every drop of rainfall in the entire area of the watershed. They built a series of bunds to prevent water runoff, ridges along farmland with trenches to collect water, and soak pits and absorption trenches to improve groundwater recharging. Elsewhere, because the watershed covered next-door Waghrul and Nandapur as well, they built gabion structures and gully plugs to hold water and further aid groundwater recharge. They studied the areas where natural rainwater streams existed and built earthen embankments across them to stop or slow down water. 'Not a drop of water is wasted,' says Kshirsagar about the 1,888-hectare revamped watershed area.

The efforts, quite literally, bore fruit. With better water availability, Kadwanchi's farmers began to grow ginger, chillies and grapes. Slowly, they added pomegranate orchards. During the rabi season, the dryland favourite sorghum is also harvested. 'We showed the way, and then others in Jalna set up vineyards too,' says Kshirsgar. Kadwanchi's residents invested nearly ₹25,000 per acre in drip irrigation systems. Kshirsagar installed a rain gauge on his farm, a useful gadget to make real-time estimates regarding the village's water budget. Everyone in Kadwanchi has micro-irrigation systems, and scores of farm ponds dot the village – square structures lined with plastic sheets to store rainwater. These ponds right next to the farms collect water through the monsoon that is then used in subsequent months.

Studied closely, it is easy to see that the reason for Kadwanchi's success is not just scientific watershed management or rainwater harvesting. In reality, what Kadwanchi's people learnt was how to manage risk and invest time and money on systems that would defend against a truant monsoon.

To some extent, the Maharashtra government's Jalyukt Shivar Abhiyan, rolled out in 2014 with a stated aim of thousands of water-scarcity-free villages by 2019, attempted to meet the same objective. The backdrop against which the scheme was conceptualized included a poor monsoon in 2011, severe drought in 2012 and worsened water scarcity as the 2014 drought dragged on. That summer, at least 188 blocks across the state had been recorded to have suffered groundwater depletion by more than two metres from a five-year average. The Jalyukt Shivar Abhiyan would be a multi-department campaign, involving a wide range of schemes such as rainwater harvesting; building water structures; widening or deepening existing water structures such as percolation tanks, storage tanks and old tanks built several decades or even centuries ago; recharging groundwater; linking small streams or nullahs; and, above all, watershed development works. A convergence system was devised so that various government department funds, corporate social responsibility funds, private agencies' assistance and people's participation could all be utilized simultaneously. The total expenditure, through various funding sources, was sizeable: ₹3,900 crore in 2015–16, ₹3,019 crore in 2016–17, ₹1,374 crore in 2017–18 and ₹627 crore in 2018–19.

The scheme got off to an enthusiastic start, but by 2015 experts had begun to question the lack of scientific planning in the digging and trenching works. Then, when the state government decided to release water from Gangapur Dam on the Godavari for a '*shahi snan*', or holy dip, by lakhs of devotees at the Kumbh Mela that year, Aurangabad-based Prof. H.M. Desarda, an economist and a former member of the state's planning board, filed a public interest litigation (PIL) challenging the release of the dam water. This was PIL number 154 of 2015. During the course of the hearings, Desarda, a seventy-six-year-old Gandhian, also submitted to the court that various policy decisions by the Maharashtra government would prove to cause irreparable damage to the state's ecology. Referring to the Jalyukt Shivar Abhiyan, he said it lacked a scientific approach.

The court appointed an expert committee to study his contentions, which opined that the project had delivered 'productive benefits' to farmers. The expert committee's report further also made some time-tested recommendations regarding micro-irrigation, river-basin water availability studies, and so on. Curiously, while Jalyukt Shivar Abhiyan works were spread across more than 16,500 villages in Maharashtra, the expert committee based its recommendations on perusal of documents and field visits to only nine villages. Of these, it found some lacunae in works completed in five. A team from the Indian Institute of Technology Bombay also assessed another six villages.

Pradeep Purandare, water sector expert and a former associate professor at the Water and Land Management Institute, Aurangabad, said the panel members had found in the nine villages they visited an incorrect outlet design in compartment bunding, improper slopes of nullahs and reduced well-water levels in the upper reaches of a watershed. In another village, the committee actually saw excessive deepening of a nullah and asked for the work to be stopped. About the IIT team's study, Purandare said the sample size was too small. In addition, he pointed out that there was no baseline survey to compare the achieved results with; besides, the team visited the sites in the post-monsoon months of September and October when water availability is relatively better.

'There were casual statements made in these reports without any elaboration or supporting documents whatsoever. For example, the report said, " ... any harm to the environment due to nullah deepening works was not observed". It was a glaring example of missing the wood for the trees,' Purandare says. Even within the small sample, the assessing teams found about 28 per cent of drainage line interventions and area-treatment interventions they visited to be 'not okay'. In two out of the six villages, the 'ridge to valley' concept was not followed. This is the central and most primary rule of watershed development programmes, to begin at the top of the ridge and conserve every drop of water there before progressively coming down the slope while reducing surface run-off and water velocity as much as possible to make for the most efficient use of water, and longevity of conservation structures lower down the slope.

According to Purandare, 'The report's silence speaks volumes. It says nothing about the basis on which claims were made regarding the water storage potential created (16.82 lakh TCM) and irrigable area created (22.5 lakh hectares). It doesn't say if this irrigable area is new or additional area, or if there is an overlap with previously irrigable areas. It says nothing about the concentration and privatization of water through plastic-lined farm ponds that don't have an inlet or outlet, or the evaporation of water from these. It also says nothing about the unofficial and undeclared redistribution of water owing to Jalyukt Shivar Abhiyan works.'

Desarda, who has written extensively on village economy and agricultural policy, says the fundamental problem with the Jalyukt Shivar Abhiyan is that the watershed approach was abandoned in the hurly-burly of political point-scoring and the rush to show progress on the ground. 'The expert committee pointed out shortcomings, and the question then is how the government would correct the anomalies ex post facto in more than 16,000 villages, where the works would have been conducted as per the guidelines given in the government resolution on Jalyukt Shivar Abhiyan.' He adds, 'The project's original guidelines should have been scrapped. Without that, there was no way to implement the recommendations of the expert committee.'

Like everywhere else in the state, Marathwada's districts also saw extensive works under the Jalyukt Shivar Abhiyan, with local BJP cadre cheering them along. But 2015 was a drought year in Marathwada, and while 2016 and 2017 saw the newly constructed or deepened water structures filled to the brim, the fervour to continue the work and maintain the structures appeared to have petered out. When the 2018 drought set in, opposition parties demanded to know which villages had achieved the drought-free status promised as part of this project. There was no easy answer.

Meanwhile, in 2016, the Government of Maharashtra, following the lead of the Union government, announced a scheme called Magel Tyala Shettala, meaning 'farm pond for anyone who asks for one'. The premise of this scheme was that at least part of one crop season could be drought-proofed by helping farmers capture run-off rainwater in plastic-lined ponds located right on their farm, making the water available for use for the next few months, with little to no effort once stored. Under the scheme, a farmer with a minimum landholding of three-fifths of a hectare can seek a maximum subsidy of ₹50,000 to build a farm pond.

In its 2016 resolution announcing the details of the project, the state government called it a 'flagship scheme'. Originally meant as a water-conservation programme for irrigation in drought-prone regions, the government later extended the scheme to water-rich Konkan as well. In three years, by March 2019, Maharashtra had 1,20,439 farm ponds under this scheme, not counting the tens of thousands of farm ponds built earlier by horticulturists, for example. Marathwada's eight districts alone have over 30,000 farm ponds. The Maharashtra government's push mirrored a Union government initiative. In his budget speeches in 2016 and 2017, the then Union finance minister Arun Jaitley had declared that the Centre had set a target of five lakh farm ponds to be built under the Mahatma Gandhi National Rural Employment Guarantee Act (MGNREGA) each year, and called it a significant step in 'drought-proofing' villages.

In fact, farm ponds began to be popularized around 2011–12 under the National Horticulture Mission or Mission for Integrated Development of Horticulture. The subsidy then was limited to 50 per cent of the cost, with a ceiling of ₹75,000 per beneficiary in the plains and ₹90,000 in hilly areas. If the pond lining, with heavy-duty plastic sheets of 300 microns thickness, was skipped, the assistance from the government would be 30 per cent less. Following its reinvention by the Centre and then rechristening by the Maharashtra government, the state scheme was meant to aid villages that had suffered crop failure, but was later extended to the entire state as farmers responded enthusiastically to the opportunity to avail of a subsidy ranging from ₹50,000 to ₹2.21 lakh.

The farm ponds are now a common feature of the rural landscape in Marathwada, their raised masonry walls visible even from the highways. In areas where horticulture is popular, such as Ambad in Jalna, there is one every couple of kilometres. They are usually empty in the summer, their blue, green or black plastic lining open to the sky. Because of the deepening water scarcity in the past few years, farmers who can afford it now have a small motor placed alongside the pond, pumping groundwater from borewells into it. In effect, water, usually seen by farmers as a common resource, is now private property. Once it is sucked out of the ground and stored in a plastic-lined pond, it belongs to one farmer alone. As farmers race to build the best store of water they can for the dry months or in anticipation of groundwater running out, deeper questions about ownership of a common resource are ignored. This also places the big farmers at an undue advantage, for they can use greater resources to build large ponds and install powerful pump sets. Moreover, this practice disadvantages villages or farms at the lower end of a watershed, an accusation that is incidentally often made against the ridge-located Kadwanchi village's success story as well.

But, just like the Jalyukt Shivar Abhiyan, the farm ponds themselves are not conceptually flawed; it is the policy and praxis surrounding them that are questionable. Both need an honest auditing by state agencies to analyse the expenditure on them and the benefits – to individual farmers, to a collective community and to long-term sustainability targets. Clouded by political exigency, a drought-proofing tool became a political subsidy handout, apparent in the decision to permit the ponds everywhere. Unfortunately, the administration has responded with silence to experts' warnings of a rapidly advancing groundwater tragedy on account of its unchecked extraction to fill the farm ponds.

Ultimately, the river-linking plans and the Marathwada Water Grid follow the same trajectory as big dams – very large contracts to private players, thousands of crores in expenditure, and running the risk of highly political choice-making in locating investments. Irrigation dams have simply not resolved Marathwada's water crisis over the years. The region's eleven major dams have a combined storage capacity of 5,143 million cubic metres (MCM). Over the past ten years, until 2018–19, the average total storage was 2,451 MCM, or 48 per cent of its capacity. The highest live storage achieved was in 2010 when it reached 4,024 MCM, or 78 per cent The lowest storage was recorded in 2013, after the failed monsoon of 2012, when the big dams reached a storage level of 834 MCM, or only 16 per cent. Seventy-five medium irrigation projects and 749 minor projects do not fare much better – their ten-year averages are 52 per cent and 40 per cent respectively, not inspiring data for a drought-prone region. In addition, utilization of irrigation potential created through these dams remains especially poor in Marathwada in the absence of distribution networks.

In 2013, the Kelkar Committee Report had suggested establishing a New Command Area Development Agency for all irrigation projects, not only related to water distribution but also to manage crop planning, mechanization, post-harvest technologies and value additions such as processing industries. It envisaged this agency as an autonomous body with accountability for achieving holistic socio-economic development of the command area. The suggestion was never taken up.

Ultimately, though, drought-proofing Marathwada cannot simply be about providing water from water-rich areas or damming the rivers. The truth is that while sparring with hydrological and agricultural drought will continue, the social dimensions of chronic drought call for making communities resilient and less vulnerable to climate shocks that could render ineffectual much of the capital-intensive engineering marvels being planned. Long-term depletion of groundwater will require solutions in sustainable cropping and integrated surface water and groundwater management.

The immediate responses to a drought in Maharashtra now fall into place like clockwork. Armies of tankers collect water from farther and farther away; private wells that are still flush are taken over by the government as additional sources; contractors are roped in to erect fodder camps where thousands of cattle are fed and watered; teams of government officials fan out to assess crop damage and announce state compensation for those affected; in some cases, farm electricity bills are waived; matriculation exam fees are waived; and, rarely, a water train is dispatched from a few hundred kilometres away, giving urban India the first sense of the deepening tragedy. But the discussion is yet to move from crisis response to managing risk, to investing big capital in mitigating the vulnerability of a people to chronic drought.

The best judges of their own conditions, Marathwada's people confirm that their top priority is sustainable sources of income through agriculture-allied activities. Wherever this has been possible, communities have emerged successful, like the women's self-help groups in Latur and Osmanabad that focus on organic farming, correct marketing and side businesses, or even the Kadwanchi village watershed management model in Jalna, which is a good example of decentralized and holistic planning for water harvesting and crop selection. But overcoming decades' worth of depleting groundwater reserves and a century's worth of slow damage to aquifers is not achievable within the span of one or two government terms. Political horizons are short, and band-aid provisions have more political capital. A long-term resolution of Marathwada's cyclical drought and its people's dependency on state aid is, sadly, not in sight.

A TIME TO LISTEN

~

W E CALLED THEM Khalistanis, separatists. But that didn't stick, so we tried other slurs to discredit the tens of thousands of farmers who have laid a desperate siege to the national capital since the end of November 2020. Surely, they must be backed by Pakistan or their unrest fomented by Maoists. At the very least, their dim view of the three farm laws they are opposing so vehemently must be coloured by the Opposition.

This historic dharna by farmers pertains to three laws enacted in September 2020 – the Farmers' Produce Trade and Commerce (Promotion and Facilitation) Act to allow the sale and purchase of farm goods anywhere, including outside the current state-regulated market system; the Farmers (Empowerment and Protection) Agreement on Price Assurance and Farm Services Act, which will promote contract farming for pre-agreed produce and prices; and the Essential Commodities (Amendment) Act, which ends stock-holding limits on, and thereby eases movement of, various agricultural commodities by removing them from the list of essential commodities. While the government, which hastily pushed the bills through Parliament during the Covid-19 pandemic, has called these laws historic and reformative, farmer groups have sought their repeal.

In Mumbai, I followed these news reports of independent India's largest ever agitation with growing incredulity, then outrage, and now gloom as there is no end in sight to the siege. The farmers' accounts of their dependence on state procurement systems and their well-evidenced, reasoned mistrust of large corporates emerging as significant stakeholders in the country's agricultural markets could not be more familiar to me. I have never spoken to a farmer from Punjab or Haryana or Rajasthan, but their narratives follow the same trajectory as those I continue to hear from Maharashtra's long-suffering agrarian class. They are talking about poor price realizations, their inability to time the market or even to understand how a globalized world order affects the price of okra or oatmeal. The smiles and songs that lend their struggle a resolute edge also make them and my Marathwada friends kindred spirits.

It is inconceivable to me how the rest of urban India does not see this. We have had a string of farmer agitations across India for a little over three years now. In June, when the laws now being opposed were promulgated as ordinances, Hansraj Wadghule-Patil, a farmer leader in Nashik, told me every farmer would welcome steps to reform the *mandi* system, but the laws had left room for profiteering by large corporates at the cost of small and marginal farmers, as well as for sudden government decisions that would render the reform worthless. As for contract farming, he told me of several instances of farmers cheated by bogus traders who bought produce on credit. The previous regime in Maharashtra had already delinked agricultural goods marketing from the *mandis* to a big extent, but the farmer community had experienced loopholes ranging from absent safeguards on traders' bonafides outside a *mandi* to lack of understanding of contract legalese.

A retired top bureaucrat from Maharashtra told me the new ordinances were attempting to change the rules of the game but not the players. For farmers to enjoy genuine choice of multiple markets, including direct sales to urban consumers or to large corporates via contracts, a laundry list of services would have to be created in every sector – seeds, fertilizer, labour, quality and grading of produce, storage, certification, marketing and so on. The laws do not spell these out, nor the impact on farmers who migrate to a new regime in the absence of these.

In September, I spoke to Wadghule-Patil again. The government had imposed a sudden ban on onion exports, and the farmers of Nashik, a major onion hub for the country, were bewildered. 'I told you so,' he said. 'Opaque government decisions will undermine their own claims of a transparent system.'

As I write this, onion prices in Lasalgaon have fallen 50 per cent over the past five weeks as red onions from the new harvest and older, uncleared produce from storage flood the market. Traders as well as farmers are repeating their pleas to lift the export ban. Farmer leaders and farmers in Maharashtra ask me a question that resonates: 'How, then, do we trust government assurances on doubling farmer incomes?'

From Nashik to Delhi, from Marathwada's flooded fields and losses this La Nina year to the unpaid dues of Uttar Pradesh's sugarcane farmers, the message from the fields is that Indian agriculture is broken. And farmers, landless labourers, tenant farmers and women farmers have all described the fault lines in great detail. And yet, in a world more connected than ever, the two Indias couldn't be farther apart, or more mistrustful of each other, unheeding and unseeing.

How did it come to be that a vast number of Indians believe Indian farmers to be lazy, blame-worthy for their own pauperization? How do we imagine that they are all enjoying tax-free incomes? Why does global data never accompany talk of state farm subsidies? The problem is not of information – there is plenty available.

Urban India's otherization of the Indian farmer likely stems from the latter having become something of a mythical creature, one that visits cities occasionally with his callused feet in chappals and his cataract-clouded eyes still looking skyward for the rains before returning to the village to don his superhero cape and produce a bumper crop, keep prices low and stay out of sight. And for the absent intelligent discourse on Indian farming among the average urban Indian, I feel the weight of failure as a journalist.

If there is one ray of hope in Delhi's current siege, it is not that a broken system can find a quick fix but that the rest of India will be able to see and meet India's farmers, hear their accounts, view their long struggle in the context of their life's work, not just vis-à-vis the three new laws and their blockade. The current recriminations do not present the wider story of the Indian farmers trapped in lifetimes of cyclical losses.

My humble hope with *Landscapes of Loss* was to do just that: to introduce readers to real people, their very real everyday struggles, their occasional triumphs and their unremitting agitations requesting justice. The countryside has spoken, rather eloquently. I hope urban India finds it in its heart to listen.

Kavitha Iyer
December 2020

SELECT BIBLIOGRAPHY

~

One: Introduction

Swaminathan Commission reports, Prof. M.S. Swaminathan, chair of the National Commission on Farmers that submitted to the central government five reports between December 2004 and October 2006, https://ruralindiaonline.org/articles/all-reports-by-the-swaminathan-commission/

Economic surveys of Maharashtra, 2012–2020, https://mahades.maharashtra.gov.in/publications.do?pubId=ESM

Economic Survey of India, 2019–20, https://www.indiabudget.gov.in/economicsurvey/

'Zimbabwe's drought-hit Bulawayo limits tap water to just a day a week', *Reuters*, 15 May 2020, https://in.reuters.com/article/us-zimbabwe-water/zimbabwes-drought-hit-bulawayo-limits-tap-water-to-just-a-day-a-week-idINKBN22R2ET

'US Drought Could Last A Century As We Now Enter A Megadrought, Study Finds', Trevor Nace, *Forbes*, 20 April 2020, https://www.forbes.com/sites/trevornace/2020/04/20/us-drought-could-last-a-century-as-we-now-enter-a-megadrought-study-finds/#30aa2c582e00

'Increased drought severity tracks warming in the United States' largest river basin', Justin T. Martin, Gregory T. Pederson and others, Proceedings of the National Academy of Sciences, May 2020, https://www.pnas.org/content/early/2020/05/05/1916208117

'Parched conditions in Germany again', NASA statement, 30 April 2020, https://earthobservatory.nasa.gov/images/146647/parched-conditions-in-germany-again

Two: Drought After Drought

Mann Deshi Foundation Report: 'My Livestock, My Wealth: Lessons from The Mann Deshi 2019 Cattle Camp', June 2019, http://manndeshifoundation.org/wp-content/uploads/2019/06/MD_Cattle-Camp-Report-2019-English.pdf

'Cattle feed to tankers: Why it's hard for relief to reach and be monitored', Kavitha Iyer, *The Indian Express*, 17 March 2016, https://indianexpress.com/article/india/india-news-india/cattle-feed-to-tankers-why-its-hard-for-relief-to-reach-and-be-monitored/

'50% of Maharashtra fodder camps are located in Marathwada', Prasad Joshi, *Times of India*, 31 March 2019, https://timesofindia.indiatimes.com/city/aurangabad/50-of-state-fodder-camps-are-located-in-marathwada/articleshow/69587085.cms

Report of the Fact-Finding Committee for Survey of Scarcity Areas in Maharashtra, S.E. Sukthankar, Government of Maharashtra, 1973

Report of the Internal Working Group to Review Agricultural Credit, RBI, September 2019

Maharashtra Pradesh Congress Committee statement on Jan Sangharsh Yatra, 23 October 2018

'He sits in AC cabin, how will CM know about drought: Ashok Chavan', *Free Press Journal*, 31 October 2018, https://www.freepressjournal.in/cmcm/he-sits-in-ac-cabin-how-will-the-cm-know-about-drought-ashok-chavan

'Maharashtra farm loan waiver to cost ₹45,000–51,000 crore: Report', PTI, 6 January 2020, https://economictimes.indiatimes.com/news/politics-and-nation/maharashtra-farm-loan-waiver-to-cost-rs-45000-51000-crore-report/articleshow/73127579.cms?from=mdr

Three: The Ever Deeper Search for Water

Asian Water Development Outlook, ADB, August 2016, https://www.adb.org/publications/asian-water-development-outlook-2016

Annual Report of the Central Ground Water Board, Ministry of Water Resources, River Development & Ganga Rejuvenation, Government of India, 2016–17, http://cgwb.gov.in/Annual-Reports/Final_Annual_Report_CGWB_2016-17_%20 03.01.2019.pdf

Report of the High-Level Committee on Balanced Regional Development Issues in Maharashtra, Dr Vijay Kelkar, Government of Maharashtra, Planning Department, October 2013, https://mahasdb.maharashtra.gov.in/docs/pdf/kcr_english_23122014.pdf

'In drought-hit Marathwada, villagers depend on tankers for water, farmers cut down fruit trees', Meena Menon, *Scroll*, 8 April 2019, https://scroll.in/article/918908/in-drought-hit-marathwada-villagers-depend-on-tankers-for-water-farmers-cut-down-fruit-trees

'Groundwater levels dip in 46 Marathwada talukas', Prasad Joshi, *Times of India*, 2 November 2019, https://timesofindia.indiatimes.com/city/aurangabad/groundwater-levels-dip-in-46-marathwada-talukas/articleshow/71860864.cms

Tourism Potential in Aurangabad: With Ajanta Ellora and Daulatabad, Dulari Qureshi, Bharatiya Kala Prakashan, 1999

World Bank project appraisal report on Atal Bhujal Yojana – National Groundwater Management Improvement Program, http://documents.worldbank.org/curated/en/697581528428694246/India-Atal-Bhujal-Yojana-ABHY-National-Groundwater-Management-Improvement-Program

Four: The Agonies of a Hardy People

Government of Maharashtra Resolution on Prerana Prakalp, https://www.maharashtra.gov.in/Site/Upload/Government%20Resolutions/English/20150915154 9221417.pdf

'Farmer suicides continue in Marathwada even as politicians promise sops ahead of polls', Radheshyam Jadhav, *The Hindu BusinessLine*, 16 October 2019, https://www.thehindubusinessline.com/news/farmer-suicides-continue-in-marathwada-even-as-politicians-promise-sops-ahead-of-polls/article29709429.ece

Nationalist Congress Party politician Supriya Sule's speech in the Rajya Sabha, during discussion on drought in eleven states, 10 May 2016, https://www.youtube.com/watch?v=fP-Hcn6daNk

'After girl commits suicide, Maharashtra waives bus pass fee for students in Marathwada', PTI, 27 October 2015, https://indianexpress.com/article/india/india-news-india/after-girl-commits-suicide-full-bus-pass-waiver-for-students-in-marathwada/

'Agriculture is Injurious to Health: Why Marathwada's Woes Continue', Atul Deulgaonkar, *Economic and Political Weekly*, 30 April 2016, https://www.epw.in/journal/2016/18/reports-states/agriculture-injurious-health.html

Five: On the Road

'Maharashtra Government sets up welfare board for sugarcane workers', ENS, 23 September 2019, https://indianexpress.com/article/cities/mumbai/maharashtra-government-sets-up-welfare-board-for-sugarcane-workers-6019469/

'Maharashtra: 61 sugar factories told to extend EPF benefits to cane labourers', Vishwas Waghmode, *The Indian Express*, 26 December 2017, https://indianexpress.com/article/india/maharashtra-61-sugar-factories-told-to-extend-epf-benefits-to-cane-labourers-4998641/

'Sugar factories federation writes to Centre, contests order', Vishwas Waghmode, *The Indian Express*, 18 June 2018, https://indianexpress.com/article/cities/mumbai/maharashtra-sugar-factories-federation-writes-to-centre-contests-order-5221905/

'Hysterectomies in Beed district raise questions for India', Patralekha Chatterjee, *The Lancet*, 20 July 2019, https://www.thelancet.com/journals/lancet/article/PIIS0140-6736(19)31669-1/fulltext

'At 36%, hysterectomies in Beed district "unusually" higher compared to state, India', ENS, 13 June 2019, https://indianexpress.com/article/cities/mumbai/maharashtra-hysterectomies-in-beed-district-unusually-higher-compared-to-state-india-5777912/

Bharatiya Janata Party politician Pankaja Munde speaks to TV channels, 20 June 2019, https://www.youtube.com/watch?v=slpI_HrtM4g

Six: A History of Exclusions

Maharashtra government resolution on regularization of encroachments on revenue lands, 28 November 1991, https://www.maharashtra.gov.in/Site/Upload/Government%20Resolutions/Ratnagiri/GENA119101.PDF

'The Marathwada Riots: A Report', Atyachar Virodhi Samiti, *Economic and Political Weekly*, 12 May 1979

'Dalit Battles For Promised Lands Rage Across India', Nihar Gokhale, *IndiaSpend*, 7 June 2019, https://www.indiaspend.com/dalit-battles-for-promised-lands-rage-across-india/

'Want to be a tribal? Welcome to Kinwat', Kanchan Srivastava, *DNA*, 2 April 2016, https://www.dnaindia.com/india/report-want-to-be-a-tribal-welcome-to-kinwat-2197095

Seven: Fleeing Home, Finding Homelessness

'As jobs dry up, migration on the rise in Marathwada', Sharad Vyas, *The Hindu*, 28 April 2016, https://www.thehindu.com/news/national/other-states/As-jobs-dry-up-migration-on-the-rise-in-Marathwada/article14260719.ece

'Parched villages begin to empty out in drought-hit Marathwada', Kavitha Iyer, *The Indian Express*, 21 November 2018, https://indianexpress.com/article/india/parched-villages-begin-to-empty-out-in-drought-hit-maharashtra-5456398/

'No water, no work: Why drought migrants in Mumbai are reluctant to go home', Mridula Chari, *Scroll*, 31 May 2016, https://scroll.in/article/809010/no-water-no-work-why-drought-migrants-in-mumbai-are-reluctant-to-go-home

Eight: Cropping a Conundrum

'Price Policy for Kharif Crops: The Marketing Season 2015–16', Commission for Agricultural Costs and Prices, Department of Agriculture and Cooperation, Ministry of Agriculture, Government of India, http://cacp.dacnet.nic.in/ViewQuestionare.aspx?Input=2&DocId=1&PageId=39&KeyId=547&utm_source=VKD&utm_medium=website&utm_campaign=e-letter&utm_content=VKD

Sugarcane price policy reports, Commission for Agricultural Costs and Prices, Ministry of Agriculture and Farmer Welfare, Government of India, http://cacp.dacnet.nic.in/KeyBullets.aspx?pid=41

Water Productivity Mapping of Major Indian Crops, NABARD and ICRIER, 2018, https://www.nabard.org/auth/writereaddata/tender/1806181128Water%20Productivity%20Mapping%20of%20Major%20Indian%20Crops,%20Web%20Version%20(Low%20Resolution%20PDF).pdf

MWRRA prohibition order for borewell more than sixty meters depth in eighty notified watersheds, 31 July 2015, https://mwrra.org/wp-content/uploads/2019/02/Prohibition-Order.pdf

'Don't allow more sugar mills in Marathwada, says Chavan', Shubhangi Khapre, *The Indian Express*, 22 August 2015, https://indianexpress.com/article/cities/mumbai/dont-allow-more-sugar-mills-in-marathwada-says-chavan/

'Water and sugarcane crushing in Maharashtra: In search of sustainability', Parineeta Dandekar, South Asia Network on Dams, Rivers and People, 15 October 2015, https://sandrp.in/2015/10/15/water-and-sugarcane-crushing-in-maharashtra-in-search-of-sustainability/

'Explained: How oil price crash impacts sugar, what it means for India', Partha Sarathi Biswas, *The Indian Express*, 26 April 2020, https://indianexpress.com/article/explained/how-oil-price-crash-impacts-sugar-what-it-means-for-india-6376389/

'Maharashtra sugar millers aim to capitalise on global prices', Partha Sarathi Biswas, *The Indian Express*, 13 February 2020, https://indianexpress.com/article/india/maharashtra-sugar-millers-aim-to-capitalise-on-global-prices-6266575/

'Maharashtra government drops case against sugar baron', Sandeep Ashar, *The Indian Express*, 19 October 2017, https://indianexpress.com/article/india/maharashtra-government-drops-case-against-sugar-baron-ratnakar-gutte-4897521/

'The story of the dead Indian farmers who applied for loans', Parth M.N., *Al Jazeera*, 19 February 2020, https://www.aljazeera.com/indepth/features/story-dead-indian-farmers-applied-loans-200218072636010.html

Bail application 956 of 2019 in the Bombay High Court, Aurangabad Bench, https://indiankanoon.org/doc/75386319/

Bombay High Court order in bail application 956 of 2019 with Criminal Application 2866 of 2019 in Gutte Vs State of Maharashtra, https://bombayhighcourt.nic.in/generatenewauth.php?bhcpar=cGF0aD0uL3dyaXRlcmVhZGRhdGEvZGF0YS9hdXJjcmltaW5hbC8yMDE5LyZmbmFtZT1BUFBMTjI4NjYxOTI2MDkxOS5wZGYmc21mbGFnPU4mcmp1ZGRhdGU9JnVwbG9hZG9PTI2LzA5LzIwMTkmc3Bhc3NwaHJhc2U9MDkxMTAxMTUyOTEw

Nine: Resistance by the Other Half

Maharashtra government resolution on transferring land titles to the names of farm widows, 2019, https://www.maharashtra.gov.in/Site/Upload/Government%20Resolutions/English/201906181503332119.pdf

'Gender Disaggregated Data on Land Ownership towards Promoting Land Rights for Women', position papers by Makaam (Mahila Kisan Adhikaar Manch), 2017 for consultation with National Commission for Women and UN Women, https://drive.google.com/file/d/1YuQMTUwreXh_yGwC-ZDGK2PYhCNfPS4q/view https://drive.google.com/file/d/1VqxcWmojukJjwd24Nu8TdOvtEH8nAVCP/view

'Differential Dependence in Early and Later Cases: Widows of Farmer Suicide Victims in Vidarbha', Kota Neelima, *Economic and Political Weekly*, 30 June 2018, https://www.epw.in/journal/2018/26-27/review-rural-affairs/widows-farmer-suicide-victims-vidarbha.html

'Public Health System is Failing Women Farmers', Nitin Jadhav, Bhausaheb Aher and Deepali Sudhindra, *Economic and Political Weekly*, 9 March 2019, https://www.epw.in/journal/2019/10/commentary/public-health-system-failing-women-farmers.html

'Ghar Doghaanche Abhiyan: Joint ownership of housing by husband and wife in Maharashtra', *Social Sector Service Delivery: Good Practices Resource Book*, Niti Aayog, Government of India, 2015, https://www.niti.gov.in/niti/writereaddata/files/bestpractices/Ghar%20Doghaanche%20Abhiyan%20Joint%20ownership%20of%20housing%20by%20husband%20and%20wife%20in%20Maharashtra.pdf

Ten: The Impossible Drought-proofing Dream

'Summary of intra-state link proposals for which PFRs completed', NWDA report, http://nwda.gov.in/upload/uploadfiles/files/aa(12).pdf

'In Marathwada, government plants ₹16,000-crore grid for piped water', Kavitha Iyer, *The Indian Express*, 17 August 2019, https://indianexpress.com/article/india/maharahstra-in-marathwada-govt-plans-rs-16000-crore-grid-for-piped-water-5911652/

Let There Be Water: Israel's Solution for a Water-Starved World, Seth M. Siegel, Tomas Dunne Books, September 2015

'Marathwada's drought will be history after Water Grid, says Fadnavis', PTI, 27 August 2019, https://www.ndtv.com/india-news/marathwada-drought-will-be-history-after-water-grid-devendra-fadnavis-2091255

'Maharashtra: Minister says ₹4,700 crore allocated for Partur, "maximum among all constituencies"', Kavitha Iyer, *The Indian Express*, 20 October 2019, https://indianexpress.com/elections/maharashtra-minister-says-rs-4700-crore-allocated-for-partur-maximum-among-all-constituencies-6078256/

HC order in Hiralal Motilal Desarda v. State of Maharashtra, Public Interest Litigation No 154 of 2015, https://bombayhighcourt.nic.in/generatenewauth.php?bhcpar=cGF0aD0uL3dyaXRlcmVhZGRhdGEvZGF0YS9qdWRnZW1lbnRzLzIwMTkvJmZuYW1lPUNQSUwyNTMyMTE1LnBkZiZzbWZsYWc9TiZyanVkZGF0ZT0mdXBsb2FkZHQ9MTgvMDIvMjAxOSZzcGFzc3BocmFzZT0xMTA1MjAwMTUxMzI=

Maharashtra government report on farm ponds, November 2019, https://egs.mahaonline.gov.in/PDF/Daily_Report_Farm_pond.pdf

Maharashtra government scheme for subsidies in building farm pond for storage of water on farms, 10 October 2016, https://egs.mahaonline.gov.in/PDF/Shetatle.pdf

'Problematic Uses and Practices of Farm Ponds in Maharashtra', Eshwer Kale, *Economic and Political Weekly*, 21 January 2017, https://wotr.org/system/files/articles/Problematic%20Uses%20and%20Practices%20of%20Farm%20Ponds%20in%20Maharashtra_Eshwer_EPW%20Article.pdf

ACKNOWLEDGEMENTS

~

JUST ABOUT EVERYTHING I was able to learn about Maharashtra and Marathwada, droughts and rural distress – in fact, whatever little I know of reporting and fact-checking – I owe to my years at the *Indian Express*. It is to this family that I hold the deepest gratitude, including my former editors Raj Kamal Jha and Unni Rajen Shanker. My former Mumbai editor Shaji Vikraman generously shared his wisdom and knowledge through the years that he oversaw my reporting and also later when I waded into the unfamiliar territory of book-length reportage.

I am deeply grateful to P. Sainath for the foreword. It is an unparalleled honour.

In the eight districts of Marathwada, I have hundreds of friends, fellow journalists, activists, doctors, ASHA workers, political workers and strangers to thank. Some cannot go unnamed.

In Beed, senior journalist Deepak Deshpande of Kaij, who first introduced me to reporting from the region's fields; Krishna

Khedkar and others of Chaklamba village in Georai, who treat me as one of their own; journalist Amol Jadhav of Nandurghat, who has always been just a call away for latest updates from the district; and two paragons of service and sacrifice, Datta Bargaje of Infant India and Deepak Nagargoje of Shantivan in Arvi.

On various occasions, complete strangers came forward to help me. I would like to especially mention Sudarshan Dalwe of Dharur, who stopped on one of the few nights when it rained in the district in August 2019 to check on a single woman stranded on the highway in the dead of night. I was parked on the pitch-black highway, thirty kilometres outside Beed city, a mild shower and dust clouding my vision, my windscreen wipers not working. Sudarshan; his sister, Rohini; and her husband, Krishna Raut, not only stopped to check on me but spent hours getting me help at midnight, and sent me a magic mechanic early the next morning. Miracles do happen, because there are still people like Sudarshan, Rohini and Krishna.

Thanks are also due to the management and staff at Hotel Anvita, Beed, who are now used to seeing me arrive at odd hours, dusty and sunburnt. Their warmth has been unexpected.

In Aurangabad, Prof. H.M. Desarda has guided me for years; water sector expert Pradeep Purandare has explained even elementary things with great patience; Dr Prasanna Patil and others at Hedgewar Hospital shared their experiences of operating distress-mitigation programmes.

In Jalna, Pandit Wasre at the Krishi Vigyan Kendra, Vishnu Piwal and Usha Chavan of Ramnagar; and journalists Nitesh Mahajan and Laxman Solunke were always willing to share their knowledge.

In Parbhani, Rajan Kshirsagar of the CPI has worked for years on working-class struggles amid a drought and was quick to share his insights; Sudhakar Kshirsagar of Sankalp gave me important lessons on the impact of seasonal migration on children.

Prema Gopalan of SSP helped me contact dozens of women farmers, besides sharing valuable insights into sustainability and gendered rebuilding of communities in distress. In Osmanabad, Naseem Shaikh and Godavari Kshirsagar of SSP spent hours discussing the changing role of women during the years of drought; Vishwanath Todkar of Paryay introduced me to the history of the region's land rights movement; Archana Bhosale of Salgara Devti village will inspire me for years to come with her incredible personal struggle. In Hingoli, Maruti Korde, who works with the Prahar Janshakti Party, introduced me to an area I might have overlooked and also helped me find transportation late one night; in Latur, residents of Jalkot taluka, where I spent many days trying to understand groundwater depletion.

In Nanded, I owe thanks to MP Prataprao Patil Chikhalikar and his gracious daughter, Pranita. Govardhan Munde of Kinwat helped me reach and explore this remote part of the region.

For willingly sharing with me his reportage of rural Maharashtra, and his widely respected byline on many occasions, I'm grateful to Partha Sarathi Biswas at the Pune edition of *Indian Express*. I'm grateful to Amit Chakravarty and Nirmal Harindran, intrepid photojournalists and my fellow travellers, for their incredible eye for beauty, drama and melancholy.

My friend Sharad Gupta came good when I suddenly demanded overnight edits of many raw chapters. To friend and colleague Mayura Janwalkar, thank you for that fantastic translation of a Marathi verse, and for coffee breaks and rambling conversations. Another friend and colleague Anushree Mazumdar gave me honest and useful feedback on one chapter, but also put her blind faith in the book overall. Sandeep Ashar, a champion investigative reporter and writer, was a reassuring sounding board, always generous with his time, his network of official sources of information and also his heartfelt encouragement.

At various points, I have badgered colleagues at the Mumbai offices of the *Indian Express* to help with data and phone numbers, and they were unfailingly tolerant and indulgent. I would like to especially mention Vishwas Waghmode, whose insights into rural Maharashtra's life and politics were always helpful.

My editor at HarperCollins India, Siddhesh Inamdar, has been a source of support through the months of delay brought on by the Covid-19 pandemic, and a calm voice of reason. I have much to learn from him. I'm also grateful to my clear-headed copy editor Suchismita Ukil at HarperCollins for lending her crisp and consistent style to the manuscript.

Though we are all mostly what we do, there is a small world I occupy outside my work, little safe havens to retreat into. These loving, nurturing humans are my personal heroes. Thank you, Zarvan Mistry, for being my rock at Ascend Fitness. Thank you, Kapil Chandni, for our 'whisky *pe vichaar*' sessions. Vivek Gilani, your insightful ideas on how to present the wider questions were invaluable.

My parents, Padma and Ramchandran Iyer, have stoically suffered my silences and absences for years. I hope this book will be worthy of them. My sister, Manisha – my first friend, best guide, loudest supporter and brightest light of our home – hers is the opinion that counts the most. It always has.

And finally, the warm-hearted people of Marathwada, men and women I have grown to love intensely. They have been my teachers, holding their grief with such tenderness and fortitude, fighting their battles with grace and a smile. Thank you.

ABOUT THE AUTHOR

~

FOR TWENTY YEARS, Kavitha Iyer's work as a journalist has revolved around recounting the stories of those on the margins, from slum dwellers in the financial capital of India to indigenous farmers in remote villages. She has written extensively on India's farm crisis, land rights, land reform, farmer suicides, distress migration and urbanization. In 2020, she was selected for a grant from the Earth Journalism Network to report on the post-Covid world of India's Adivasis participating in land rights movements. She was also a grantee of the Thakur Foundation, USA, for a project documenting the struggle of rural healthcare to cope with the Covid-19 pandemic in the face of historical under-investment and neglect.

Kavitha is editor of *26/11: Stories of Strength*, a compilation of accounts from the victims of the Mumbai terror attacks of 2008.

Early in her career, she won a National Foundation for India (NFI) fellowship to spend a year researching and writing on Dharavi, Asia's largest slum.

A true-blue Mumbaikar, Kavitha has taught journalism and mentored several young journalists during her career as city editor and then associate editor at the *Indian Express*. In two stints spanning over seventeen years with the *Indian Express*, she travelled across Maharashtra's hinterland trying to locate the big India story in the countryside, and tracking farmer unrest, political agitations, indebtedness, hunger, malnutrition, deprivation, cyclical crop loss, climate change and drought, and the intersection of these with policy.

Now an independent journalist, she spends her free time driving solo in rural Maharashtra, breaking bread with strangers and enjoying the sense of being quite lost.